BRICKLAYING

LEVEL 3 DIPLOMA

JOHN CARRUTHERS
IAN COOTE

Leeds College of Building

Nelson Thornes

Published in 2013 by:
Nelson Thornes Ltd
Delta Place
27 Bath Road
CHELTENHAM
GL53 7TH
United Kingdom

13 14 15 16 17 / 10 9 8 7 6 5 4 3 2 1

A catalogue record for this book is available from the British Library
ISBN 978 1 4085 2124 3

Cover photograph: DNY59/iStockphoto

Page make-up by GreenGate Publishing Services, Tonbridge, Kent

Printed in Croatia by Zrinski

Note to learners and tutors

This book clearly states that a risk assessment should be undertaken and the correct PPE worn for the particular activities before any practical activity is carried out. Risk assessments were carried out before photographs for this book were taken and the models are wearing the PPE deemed appropriate for the activity and situation. This was correct at the time of going to print. Colleges may prefer that their learners wear additional items of PPE not featured in the photographs in this book and should instruct learners to do so in the standard risk assessments they hold for activities undertaken by their learners. Learners should follow the standard risk assessments provided by their college for each activity they undertake which will determine the PPE they wear.

CONTENTS

INTRODUCTION

About this book

This book has been written for the Cskills Awards Level 3 Diploma in Bricklaying. It covers all the units of the qualification, so you can feel confident that your book fully covers the requirements of your course.

This book contains a number of features to help you acquire the knowledge you need. It also demonstrates the practical skills you will need to master to successfully complete your qualification. We've included additional features to show how the skills and knowledge can be applied to the workplace, as well as tips and advice on how you can improve your chances of gaining employment.

The features include:

* chapter openers which list the learning outcomes you must achieve in each unit

* key terms that provide explanations of important terminology that you will need to know and understand

* Did you know? margin notes to provide key facts that are helpful to your learning

* practical tips to explain facts or skills to remember when undertaking practical tasks

* Reed tips to offer advice about work, building your CV and how to apply the skills and knowledge you have learnt in the workplace

* case studies that are based on real tradespeople who have undertaken apprenticeships and explain why the skills and knowledge you learn with your training provider are useful in the workplace

* practical tasks that provide step-by-step directions and illustrations for a range of projects you may do during your course

* Test yourself multiple choice questions that appear at the end of each unit to give you the chance to revise what you have learnt and to practise your assessment (your tutor will give you the answers to these questions).

Further support for this book can be found at our website,

www.planetvocational.com/subjects/build

CONTRIBUTORS TO THIS BOOK

Reed Property & Construction

Reed Property & Construction specialises in placing staff at all levels, in both temporary and permanent positions, across the complete lifecycle of the construction process. Our consultants work with most major construction companies in the UK and our clients are involved with the design, build and maintenance of infrastructure projects throughout the UK.

Expert help

As a leading recruitment consultancy for mid–senior level construction staff in the UK, Reed Property & Construction is ideally placed to advise new workers entering the sector, from building a CV to providing expertise and sharing our extensive sector knowledge with you. That's why, throughout this book, you will find helpful hints from our highly experienced consultants, all designed to help you find that first step on the construction career ladder. These tips range from advice on CV writing to interview tips and techniques, and are all linked in with the learning material in this book.

Work-related advice

Reed Property & Construction has gained insights from some of our biggest clients – leading recruiters within the industry – to help you understand the mind-set of potential employers. This includes the traits and skills that they would like to see in their new employees, why you need the skills taught in this book and how they are used on a day to day basis within their organisations.

Getting your first job

This invaluable information is not available anywhere else and is all geared towards helping you gain a position with an employer once you've completed your studies. Entry level positions are not usually offered by recruitment companies, but the advice we've provided will help you to apply for jobs in construction and hopefully gain your first position as a skilled worker.

CONTRIBUTORS TO THIS BOOK

The case studies in this book feature staff from Laing O'Rourke and South Tyneside Homes.

Laing O'Rourke is an international engineering company that constructs large-scale building projects all over the world. Originally formed from two companies, John Laing (founded in 1848) and R O'Rourke and Son (founded in 1978) joined forces in 2001.

At Laing O'Rourke, there is a strong and unique apprenticeship programme. It runs a four-year 'Apprenticeship Plus' scheme in the UK, combining formal college education with on-the-job training. Apprentices receive support and advice from mentors and experienced tradespeople, and are given the option of three different career pathways upon completion: remaining on site, continuing into a further education programme, or progressing into supervision and management.

The company prides itself on its people development, supporting educational initiatives and investing in its employees. Laing O'Rourke believes in collaboration and teamwork as a path to achieving greater success, and strives to maintain exceptionally high standards in workplace health and safety.

South Tyneside Homes was launched in 2006, and was previously part of South Tyneside Council. It now works in partnership with the council to repair and maintain 18,000 properties within the borough, including delivering parts of the Decent Homes Programme.

South Tyneside Homes believes in putting back into the community, with 90 per cent of its employees living in the borough itself. Equality and diversity, as well as health and wellbeing of staff, is a top priority, and it has achieved the Gold Status Investors in People Award.

South Tyneside Homes is committed to the development of its employees, providing opportunities for further education and training and great career paths within the company – 80 per cent of its management team started as apprentices with the company. As well as looking after its staff and their community, the company looks after the environment too, running a renewable energy scheme for council tenants in order to reduce carbon emissions and save tenants money.

The apprenticeship programme at South Tyneside Homes has been recognised nationally, having trained over 80 young people in five main trade areas over the past six years. One of the UK's Top 100 Apprenticeship Employers, it is an Ambassador on the panel of the National Apprentice Service. It has won the Large Employer of the Year Award at the National Apprenticeship Awards and several of its apprentices have been nominated for awards, including winning the Female Apprentice of the Year for the local authority.

Chapter 1

Unit CSA–L1Core01

HEALTH, SAFETY AND WELFARE IN CONSTRUCTION AND ASSOCIATED INDUSTRIES

LEARNING OUTCOMES

LO1: Know the health and safety regulations, roles and responsibilities

LO2: Know the accident and emergency procedures and how to report them

LO3: Know how to identify hazards on construction sites

LO4: Know about health and hygiene in a construction environment

LO5: Know how to handle and store materials and equipment safely

LO6: Know about basic working platforms and access equipment

LO7: Know how to work safely around electricity in a construction environment

LO8: Know how to use personal protective equipment (PPE) correctly

LO9: Know the fire and emergency procedures

LO10: Know about signs and safety notices

INTRODUCTION

The aim of this chapter is to:

* help you to source relevant safety information
* help you to use the relevant safety procedures at work.

KEY TERMS

HASAWA

– the Health and Safety at Work etc. Act outlines your and your employer's health and safety responsibilities.

COSHH

– the Control of Substances Hazardous to Health Regulations are concerned with controlling exposure to hazardous materials.

DID YOU KNOW?

In 2011 to 2012, there were 49 fatal accidents in the construction industry in the UK. (*Source* HSE, www.hse.gov.uk)

KEY TERMS

HSE

– the Health and Safety Executive, which ensures that health and safety laws are followed.

Accident book

– this is required by law under the Social Security (Claims and Payments) Regulations 1979. Even minor accidents need to be recorded by the employer. For the purposes of RIDDOR, hard copy accident books or online records of incidents are equally acceptable.

HEALTH AND SAFETY REGULATIONS, ROLES AND RESPONSIBILITIES

The construction industry can be dangerous, so keeping safe and healthy at work is very important. If you are not careful, you could injure yourself in an accident or perhaps use equipment or materials that could damage your health. Keeping safe and healthy will help ensure that you have a long and injury-free career.

Although the construction industry is much safer today than in the past, more than 2,000 people are injured and around 50 are killed on site every year. Many others suffer from long-term ill-health such as deafness, spinal damage, skin conditions or breathing problems.

Key health and safety legislation

Laws have been created in the UK to try to ensure safety at work. Ignoring the rules can mean injury or damage to health. It can also mean losing your job or being taken to court.

The two main laws are the Health and Safety at Work etc. Act **(HASAWA)** and the Control of Substances Hazardous to Health Regulations **(COSHH)**.

The Health and Safety at Work etc. Act (HASAWA) (1974)

This law applies to all working environments and to all types of worker, sub-contractor, employer and all visitors to the workplace. It places a duty on everyone to follow rules in order to ensure health, safety and welfare. Businesses must manage health and safety risks, for example by providing appropriate training and facilities. The Act also covers first aid, accidents and ill health.

Reporting of Injuries, Diseases and Dangerous Occurrences Regulations (RIDDOR) (1995)

Under RIDDOR, employers are required to report any injuries, diseases or dangerous occurrences to the **Health and Safety Executive (HSE)**. The regulations also state the need to maintain an **accident book**.

Control of Substances Hazardous to Health (COSHH) (2002)

In construction, it is common to be exposed to substances that could cause ill health. For example, you may use oil-based paints or preservatives, or work in conditions where there is dust or bacteria.

Employers need to protect their employees from the risks associated with using hazardous substances. This means assessing the risks and deciding on the necessary precautions to take.

Any control measures (things that are being done to reduce the risk of people being hurt or becoming ill) have to be introduced into the workplace and maintained; this includes monitoring an employee's exposure to harmful substances. The employer will need to carry out health checks and ensure that employees are made aware of the dangers and are supervised.

Control of Asbestos at Work Regulations (2012)

Asbestos was a popular building material in the past because it was a good insulator, had good fire protection properties and also protected metals against corrosion. Any building that was constructed before 2000 is likely to have some asbestos. It can be found in pipe insulation, boilers and ceiling tiles. There is also asbestos cement roof sheeting and there is a small amount of asbestos in decorative coatings such as Artex.

Asbestos has been linked with lung cancer, other damage to the lungs and breathing problems. The regulations require you and your employer to take care when dealing with asbestos:

* You should always assume that materials contain asbestos unless it is obvious that they do not.

* A record of the location and condition of asbestos should be kept.

* A risk assessment should be carried out if there is a chance that anyone will be exposed to asbestos.

The general advice is as follows:

* Do not remove the asbestos. It is not a hazard unless it is removed or damaged.

* Remember that not all asbestos presents the same risk. Asbestos cement is less dangerous than pipe insulation.

* Call in a specialist if you are uncertain.

Provision and Use of Work Equipment Regulations (PUWER) (1998)

PUWER concerns health and safety risks related to equipment used at work. It states that any risks arising from the use of equipment must either be prevented or controlled, and all suitable safety measures must have been taken. In addition, tools need to be:

* suitable for their intended use

* safe

REED TIP

Employers will want to know that you understand the importance of health and safety. Make sure you know the reasons for each safe working practice.

* well maintained

* used only by those who have been trained to do so.

Manual Handling Operations Regulations (1992)

These regulations try to control the risk of injury when lifting or handling bulky or heavy equipment and materials. The regulations state as follows:

* Hazardous manual handling should be avoided if possible.

* An assessment of hazardous manual handling should be made to try to find alternatives.

* You should use mechanical assistance where possible.

* The main idea is to look at how manual handling is carried out and finding safer ways of doing it.

Personal Protection at Work Regulations (PPE) (1992)

This law states that employers must provide employees with personal protective equipment **(PPE)** at work whenever there is a risk to health and safety. PPE needs to be:

* suitable for the work being done

* well maintained and replaced if damaged

* properly stored

* correctly used (which means employees need to be trained in how to use the PPE properly).

Work at Height Regulations (2005)

Whenever a person works at any height there is a risk that they could fall and injure themselves. The regulations place a duty on employers or anyone who controls the work of others. This means that they need to:

* plan and organise the work

* make sure those working at height are **competent**

* assess the risks and provide appropriate equipment

* manage work near or on fragile surfaces

* ensure equipment is inspected and maintained.

In all cases the regulations suggest that, if it is possible, work at height should be avoided. Perhaps the job could be done from ground level? If it is not possible, then equipment and other measures are needed to prevent the risk of falling. When working at height measures also need to be put in place to minimise the distance someone might fall.

KEY TERMS

PPE

– personal protective equipment can include gloves, goggles and hard hats.

Competent

– to be competent an organisation or individual must have:

* sufficient knowledge of the tasks to be undertaken and the risks involved

* the experience and ability to carry out their duties in relation to the project, to recognise their limitations and take appropriate action to prevent harm to those carrying out construction work, or those affected by the work.

(*Source* HSE)

Figure 1.1 Examples of personal protective equipment

Employer responsibilities under HASAWA

HASAWA states that employers with five or more staff need their own health and safety policy. Employers must assess any risks that may be involved in their workplace and then introduce controls to reduce these risks. These risk assessments need to be reviewed regularly.

Employers also need to supply personal protective equipment (PPE) to all employees when it is needed and to ensure that it is worn when required.

Specific employer responsibilities are outlined in Table 1.1.

Employee responsibilities under HASAWA

HASAWA states that all those operating in the workplace must aim to work in a safe way. For example, they must wear any PPE provided and look after their equipment. Employees should not be charged for PPE or any actions that the employer needs to take to ensure safety.

Specific employer responsibilities are outlined in Table 1.1. Table 1.2 identifies the key employee responsibilities.

KEY TERMS

Risk

– the likelihood that a person may be harmed if they are exposed to a hazard.

Hazard

– a potential source of harm, injury or ill-health.

Near miss

– any incident, accident or emergency that did not result in an injury but could have done so.

Employer responsibility	Explanation
Safe working environment	Where possible all potential risks and hazards should be eliminated.
Adequate staff training	When new employees begin a job their induction should cover health and safety. There should be ongoing training for existing employees on risks and control measures.
Health and safety information	Relevant information related to health and safety should be available for employees to read and have their own copies.
Risk assessment	Each task or job should be investigated and potential risks identified so that measures can be put in place. A risk assessment and method statement should be produced. The method statement will tell you how to carry out the task, what PPE to wear, equipment to use and the sequence of its use.
Supervision	A competent and experienced individual should always be available to help ensure that health and safety problems are avoided.

Table 1.1 Employer responsibilities under HASAWA

Employee responsibility	Explanation
Working safely	Employees should take care of themselves, only do work that they are competent to carry out and remove obvious hazards if they are seen.
Working in partnership with the employer	Co-operation is important and you should never interfere with or misuse any health and safety signs or equipment. You should always follow the site rules.
Reporting hazards, near misses and accidents correctly	Any health and safety problems should be reported and discussed, particularly a near miss or an actual accident.

Table 1.2 Employee responsibilities under HASAWA

Health and Safety Executive

The Health and Safety Executive (HSE) is responsible for health, safety and welfare. It carries out spot checks on different workplaces to make sure that the law is being followed.

HSE inspectors have access to all areas of a construction site and can also bring in the police. If they find a problem then they can issue an **improvement notice**. This gives the employer a limited amount of time to put things right.

In serious cases, the HSE can issue a **prohibition notice**. This means all work has to stop until the problem is dealt with. An employer, the employees or **sub-contractors** could be taken to court.

The roles and responsibilities of the HSE are outlined in Table 1.3.

Responsibility	Explanation
Enforcement	It is the HSE's responsibility to reduce work-related death, injury and ill health. It will use the law against those who put others at risk.
Legislation and advice	The HSE will use health and safety legislation to serve improvement or prohibition notices or even to prosecute those who break health and safety rules. Inspectors will provide advice either face-to-face or in writing on health and safety matters.
Inspection	The HSE will look at site conditions, standards and practices and inspect documents to make sure that businesses and individuals are complying with health and safety law.

Table 1.3 HSE roles and responsibilities

Sources of health and safety information

There is a wide variety of health and safety information. Most of it is available free of charge, while other organisations may make a charge to provide information and advice. Table 1.4 outlines the key sources of health and safety information.

Source	Types of information	Website
Health and Safety Executive (HSE)	The HSE is the primary source of work-related health and safety information. It covers all possible topics and industries.	www.hse.gov.uk
Construction Industry Training Board (CITB)	The national training organisation provides key information on legislation and site safety.	www.citb.co.uk
British Standards Institute (BSI)	Provides guidelines for risk management, PPE, fire hazards and many other health and safety-related areas.	www.bsigroup.com
Royal Society for the Prevention of Accidents (RoSPA)	Provides training, consultancy and advice on a wide range of health and safety issues that are aimed to reduce work related accidents and ill health.	www.rospa.com
Royal Society for Public Health (RSPH)	Has a range of qualifications and training programmes focusing on health and safety.	www.rsph.org.uk

Table 1.4 Health and safety information

Informing the HSE

The HSE requires the reporting of:

* deaths and injuries – any **major injury**, **over 7-day injury** or death

* occupational disease

* dangerous occurrence – a collapse, explosion, fire or collision

* gas accidents – any accidental leaks or other incident related to gas.

Enforcing guidance

Work-related injuries and illnesses affect huge numbers of people. According to the HSE, 1.1 million working people in the UK suffered from a work-related illness in 2011 to 2012. Across all industries, 173 workers were killed, 111,000 other injuries were reported and 27 million working days were lost.

The construction industry is a high risk one and, although only around 5 per cent of the working population is in construction, it accounts for 10 per cent of all major injuries and 22 per cent of fatal injuries.

The good news is that enforcing guidance on health and safety has driven down the numbers of injuries and deaths in the industry. Only 20 years ago over 120 construction workers died in workplace accidents each year. This is now reduced to fewer than 60 a year.

However, there is still more work to be done and it is vital that organisations such as the HSE continue to enforce health and safety and continue to reduce risks in the industry.

On-site safety inductions and toolbox talks

The HSE suggests that all new workers arriving on site should attend a short induction session on health and safety. It should:

* show the commitment of the company to health and safety

* explain the health and safety policy

* explain the roles individuals play in the policy

* state that each individual has a legal duty to contribute to safe working

* cover issues like excavations, work at height, electricity and fire risk

* provide a layout of the site and show evacuation routes

* identify where fire fighting equipment is located

* ensure that all employees have evidence of their skills

* stress the importance of signing in and out of the site.

DID YOU KNOW?

Toolbox talks are normally given by a supervisor and often take place on site, either during the course of a normal working day or when someone has been seen working in an unsafe way. CITB produces a book called *GT700 Toolbox Talks* which covers a range of health and safety topics, from trying a new process and using new equipment to particular hazards or work practices.

Behaviour and actions that could affect others

It is the responsibility of everyone on site not only to look after their own health and safety, but also to ensure that their actions do not put anyone else at risk.

Trying to carry out work that you are not competent to do is not only dangerous to yourself but could compromise the safety of others.

Simple actions, such as ensuring that all of your rubbish and waste is properly disposed of, will go a long way to removing hazards on site that could affect others.

Just as you should not create a hazard, ignoring an obvious one is just as dangerous. You should always obey site rules and particularly the health and safety rules. You should follow any instructions you are given.

ACCIDENT AND EMERGENCY PROCEDURES

PRACTICAL TIP

If you come across any health and safety problems you should report them so that they can be controlled.

All sites will have specific procedures for dealing with accidents and emergencies. An emergency will often mean that the site needs to be evacuated, so you should know in advance where to assemble and who to report to. The site should never be re-entered without authorisation from an individual in charge or the emergency services.

Types of emergencies

Emergencies are incidents that require immediate action. They can include:

- fires
- spillages or leaks of chemicals or other hazardous substances, such as gas
- failure of a scaffold
- collapse of a wall or trench
- a health problem
- an injury
- bombs and security alerts.

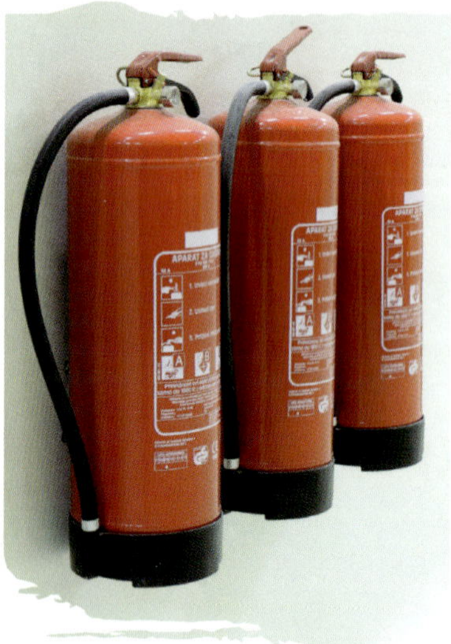

Figure 1.2 It's important that you know where your company's fire-fighting equipment is located

Legislation and reporting accidents

RIDDOR (1995) puts a duty on employers, anyone who is self-employed, or an individual in control of the work, to report any serious workplace accidents, occupational diseases or dangerous occurrences (also known as near misses).

The report has to be made by these individuals and, if it is serious enough, the responsible person may have to fill out a RIDDOR report.

Injuries, diseases and dangerous occurrences

Construction sites can be dangerous places, as we have seen. The HSE maintains a list of all possible injuries, diseases and dangerous occurrences, particularly those that need to be reported.

Injuries

There are two main classifications of injuries: minor and major. A minor injury can usually be handled by a competent first aider, although it is often a good idea to refer the individual to their doctor or to the hospital. Typical minor injuries can include:

* minor cuts   * minor burns   * exposure to fumes.

Major injuries are more dangerous and will usually require the presence of an ambulance with paramedics. Major injuries can include:

* bone fracture   * concussion

* unconsciousness   * electric shock.

Diseases

There are several different diseases and health issues that have to be reported, particularly if a doctor notifies that a disease has been diagnosed. These include:

* poisoning   * infections

* skin diseases   * occupational cancer

* lung diseases   * hand/arm vibration syndrome.

Dangerous occurrences

Even if something happens that does not result in an injury, but could easily have done so, it is classed as a dangerous occurrence. It needs to be reported immediately and then followed up by an accident report form. Dangerous occurrences can include:

* accidental release of a substance that could damage health

* anything coming into contact with overhead power lines

* an electrical problem that caused a fire or explosion

* collapse or partial collapse of scaffolding over 5 m high.

> **PRACTICAL TIP**
>
> An up-to-date list of dangerous occurrences is maintained by the Health and Safety Executive.

Recording accidents and emergencies

The Reporting of Injuries, Diseases and Dangerous Occurrences Regulations (RIDDOR) (1995) requires employers to:

* report any relevant injuries, diseases or dangerous occurrences to the Health and Safety Executive (HSE)

* keep records of incidents in a formal and organised manner (for example, in an accident book or online database).

After an accident, you may need to complete an accident report form – either in writing or online. This form may be completed by the person who was injured or the first aider.

On the accident report form you need to note down:

* the casualty's personal details, e.g. name, address, occupation

* the name of the person filling in the report form

* the details of the accident.

In addition, the person reporting the accident will need to sign the form.

On site a trained first aider will be the first individual to try and deal with the situation. In addition to trying to save life, stop the condition from getting worse and getting help, they will also record the occurrence.

On larger sites there will be a health and safety officer, who would keep records and documentation detailing any accidents and emergencies that have taken place on site. All companies should keep such records; it may be a legal requirement for them to do so under RIDDOR and it is good practice to do so in case the HSE asks to see it.

Importance of reporting accidents and near misses

Reporting incidents is not just about complying with the law or providing information for statistics. Each time an accident or near miss takes place it means lessons can be learned and future problems avoided.

The accident or near miss can alert the business or organisation to a potential problem. They can then take steps to ensure that it does not occur in the future.

Major and minor injuries and near misses

RIDDOR defines a major injury as:

* a fracture (but not to a finger, thumb or toes)

* a dislocation

* an amputation

* a loss of sight in an eye

* a chemical or hot metal burn to the eye

* a penetrating injury to the eye

* an electric shock or electric burn leading to unconsciousness and/or requiring resuscitation

* hyperthermia, heat-induced illness or unconsciousness

* asphyxia

* exposure to a harmful substance

* inhalation of a substance

* acute illness after exposure to toxins or infected materials.

A minor injury could be considered as any occurrence that does not fall into any of the above categories.

A near miss is any incident that did not actually result in an injury but which could have caused a major injury if it had done so. Non-reportable near misses are useful to record as they can help to identify potential problems. Looking at a list of near misses might show patterns for potential risk.

Accident trends

We have already seen that the HSE maintains statistics on the number and types of construction accidents. The following are among the 2011/2012 construction statistics:

* There were 49 fatalities.

* There were 5,000 occupational cancer patients.

* There were 74,000 cases of work-related ill health.

* The most common types of injury were caused by falls, although many injuries were caused by falling objects, collapses and electricity. A number of construction workers were also hurt when they slipped or tripped, or were injured while lifting heavy objects.

Accidents, emergencies and the employer

Even less serious accidents and injuries can cost a business a great deal of money. But there are other costs too:

* Poor company image – if a business does not have health and safety controls in place then it may get a reputation for not caring about its employees. The number of accidents and injuries may be far higher than average.

* Loss of production – the injured individual might have to be treated and then may need a period of time off work to recover. The loss of production can include those who have to take time out from working to help the injured person and the time of a manager or supervisor who has to deal with all the paperwork and problems.

* Insurance – each time there is an accident or injury claim against the company's insurance the premiums will go up. If there are many accidents and injuries the business may find it impossible to get insurance. It is a legal requirement for a business to have insurance so in the end that company might have to close down.

* Closure of site – if there is a serious accident or injury then the site may have to be closed while investigations take place to discover the reason, or who was responsible. This could cause serious delays and loss of income for workers and the business.

DID YOU KNOW?

RoSPA (the Royal Society for the Prevention of Accidents) uses many of the statistics from the HSE. The latest figures that RoSPA has analysed date back to 2008/2009. In that year, 1.2 million people in the UK were suffering from work-related illnesses. With fewer than 132,000 reportable injuries at work, this is believed to be around half of the real figure.

DID YOU KNOW?

An employee working in a small business broke two bones in his arm. He could not return to proper duties for eight months. He lost out on wages while he was off sick and, in total, it cost the business over £45,000.

REED TIP

On some construction sites, you may get a Health and Safety Inspector come to look round without any notice – one more reason to always be thinking about working safely.

Accident and emergency authorised personnel

Several different groups of people could be involved in dealing with accident and emergency situations. These are listed in Table 1.5.

Authorised personnel	Role
First aiders and emergency responders	These are employees on site and in the workforce who have been trained to be the first to respond to accidents and injuries. The minimum provision of an appointed person would be someone who has had basic first aid training. The appointment of a first aider is someone who has attained a higher or specific level of training. A construction site with fewer than 5 employees needs an appointed first aider. A construction site with up to 50 employees requires a trained first aider, and for bigger sites at least one trained first aider is required for every 50 people.
Supervisors and managers	These have the responsibility of managing the site and would have to organise the response and contact emergency services if necessary. They would also ensure that records of any accidents are completed and up to date and notify the HSE if required.
Health and Safety Executive	The HSE requires businesses to investigate all accidents and emergencies. The HSE may send an inspector, or even a team, to investigate and take action if the law has been broken.
Emergency services	Calling the emergency services depends on the seriousness of the accident. Paramedics will take charge of the situation if there is a serious injury and if they feel it necessary will take the individual to hospital.

Table 1.5 People who deal with accident and emergency situations

DID YOU KNOW?

The three main emergency services in the UK are: the Fire Service (for fire and rescue); the Ambulance Service (for medical emergencies); the Police (for an immediate police response). Call them on 999 only if it is an emergency.

The basic first aid kit

BS 8599 relates to first aid kits, but it is not legally binding. The contents of a first aid box will depend on an employer's assessment of their likely needs. The HSE does not have to approve the contents of a first aid box but it states that where the work involves low level hazards the minimum contents of a first aid box should be:

* a copy of its leaflet on first aid – *HSE Basic advice on first aid at work*

* 20 sterile plasters of assorted size

* 2 sterile eye pads

* 4 sterile triangular bandages

* 6 safety pins

* 2 large sterile, unmedicated wound dressings

* 6 medium-sized sterile unmedicated wound dressings

* 1 pair of disposable gloves.

The HSE also recommends that no tablets or medicines are kept in the first aid box.

Figure 1.3 A typical first aid box

What to do if you discover an accident

When an accident happens it may not only injure the person involved directly, but it may also create a hazard that could then injure others. You need to make sure that the area is safe enough for you or someone else to help the injured person. It may be necessary to turn off the electrical supply or remove obstructions to the site of the accident.

The first thing that needs to be done if there is an accident is to raise the alarm. This could mean:

* calling for the first aider

* phoning for the emergency services

* dealing with the problem yourself.

How you respond will depend on the severity of the injury.

You should follow this procedure if you need to contact the emergency services:

* Find a telephone away from the emergency.

* Dial 999.

* You may have to go through a switchboard. Carefully listen to what the operator is saying to you and try to stay calm.

* When asked, give the operator your name and location, and the name of the emergency service or services you require.

* You will then be transferred to the appropriate emergency service, who will ask you questions about the accident and its location. Answer the questions in a clear and calm way.

* Once the call is over, make sure someone is available to help direct the emergency services to the location of the accident.

IDENTIFYING HAZARDS

As we have already seen, construction sites are potentially dangerous places. The most effective way of handling health and safety on a construction site is to spot the hazards and deal with them before they can cause an accident or an injury. This begins with basic housekeeping and carrying out risk assessments. It also means having a procedure in place to report hazards so that they can be dealt with.

Good housekeeping

Work areas should always be clean and tidy. Sites that are messy, strewn with materials, equipment, wires and other hazards can prove to be very dangerous. You should:

* always work in a tidy way

* never block fire exits or emergency escape routes

* never leave nails and screws scattered around

* ensure you clean and sweep up at the end of each working day

* not block walkways

* never overfill skips or bins

* never leave food waste on site.

Risk assessments and method statements

It is a legal requirement for employers to carry out risk assessments. This covers not only those who are actually working on a particular job, but other workers in the immediate area, and others who might be affected by the work.

It is important to remember that when you are carrying out work your actions may affect the safety of other people. It is important, therefore, to know whether there are any potential hazards. Once you know what these hazards are you can do something to either prevent or reduce them as a risk. Every job has potential hazards.

There are five simple steps to carrying out a risk assessment, which are shown in Table 1.6, using the example of repointing brickwork on the front face of a dwelling.

Step	Action	Example
1	Identify hazards	The property is on a street with a narrow pavement. The damaged brickwork and loose mortar need to be removed and placed in a skip below. Scaffolding has been erected. The road is not closed to traffic.
2	Identify who is at risk	The workers repointing are at risk as they are working at height. Pedestrians and vehicles passing are at risk from the positioning of the skip and the chance that debris could fall from height.
3	What is the risk from the hazard that may cause an accident?	The risk to the workers is relatively low as they have PPE and the scaffolding has been correctly erected. The risk to those passing by is higher, as they are unaware of the work being carried out above them.
4	Measures to be taken to reduce the risk	Station someone near the skip to direct pedestrians and vehicles away from the skip while the work is being carried out. Fix a secure barrier to the edge of the scaffolding to reduce the chance of debris falling down. Lower the bricks and mortar debris using a bucket or bag into the skip and not throwing them from the scaffolding. Consider carrying out the work when there are fewer pedestrians and less traffic on the road.
5	Monitor the risk	If there are problems with the first stages of the job, you need to take steps to solve them. If necessary consider taking the debris by hand through the building after removal.

Table 1.6 A five-step risk assessment for repointing brickwork

Your employer should follow these working practices, which can help to prevent accidents or dangerous situations occurring in the workplace:

* *Risk assessments* look carefully at what could cause an individual harm and how to prevent this. This is to ensure that no one should be injured or become ill as a result of their work. Risk assessments identify how likely it is that an accident might happen and the consequences of it happening. A risk factor is worked out and control measures created to try to offset them.

* *Method statements,* however brief, should be available for every risk assessment. They summarise risk assessments and other findings to provide guidance on how the work should be carried out.

* *Permit to work systems* are used for very high risk or even potentially fatal activities. They are checklists that need to be completed before the work begins. They must be signed by a supervisor.

* *A hazard book* lists standard tasks and identifies common hazards. These are useful tools to help quickly identify hazards related to particular tasks.

Types of hazards

Typical construction accidents can include:

* fires and explosions

* slips, trips and falls

* burns, including those from chemicals

* falls from scaffolding, ladders and roofs

* electrocution

* injury from faulty machinery

* power tool accidents

* being hit by construction debris

* falling through holes in flooring.

We will look at some of the more common hazards in a little more detail.

Fires
Fires need oxygen, heat and fuel to burn. Even a spark can provide enough heat needed to start a fire, and anything flammable, such as petrol, paper or wood, provides the fuel. It may help to remember the 'triangle of fire' – heat, oxygen and fuel are all needed to make fire so remove one or more to help prevent or stop the fire.

Tripping

Leaving equipment and materials lying around can cause accidents, as can trailing cables and spilt water or oil. Some of these materials are also potential fire hazards.

Chemical spills

If the chemicals are not hazardous then they just need to be mopped up. But sometimes they do involve hazardous materials and there will be an existing plan on how to deal with them. A risk assessment will have been carried out.

Falls from height

A fall even from a low height can cause serious injuries. Precautions need to be taken when working at height to avoid permanent injury. You should also consider falls into open excavations as falls from height. All the same precautions need to be in place to prevent a fall.

Burns

Burns can be caused not only by fires and heat, but also from chemicals and solvents. Electricity and wet concrete and cement can also burn skin. PPE is often the best way to avoid these dangers. Sunburn is a common and uncomfortable form of burning and sunscreen should be made available. For example, keeping skin covered up will help to prevent sunburn. You might think a tan looks good, but it could lead to skin cancer.

Electrical

Electricity is hazardous and electric shocks can cause burns and muscle damage, and can kill.

Exposure to hazardous substances

We look at hazardous substances in more detail on pages 20–1. COSHH regulations identify hazardous substances and require them to be labelled. You should always follow the instructions when using them.

Plant and vehicles

On busy sites there is always a danger from moving vehicles and heavy plant. Although many are fitted with reversing alarms, it may not be easy to hear them over other machinery and equipment. You should always ensure you are not blocking routes or exits. Designated walkways separate site traffic and pedestrians – this includes workers who are walking around the site. Crossing points should be in place for ease of movement on site.

Reporting hazards

We have already seen that hazards have the potential to cause serious accidents and injuries. It is therefore important to report hazards and there are different methods of doing this.

The first major reason to report hazards is to prevent danger to others, whether they are other employees or visitors to the site. It is vital to prevent accidents from taking place and to quickly correct any dangerous situations.

Injuries, diseases and actual accidents all need to be reported and so do dangerous occurrences. These are incidents that do not result in an actual injury, but could easily have hurt someone.

Accidents need to be recorded in an accident book, computer database or other secure recording system, as do near misses. Again it is a legal requirement to keep appropriate records of accidents and every company will have a procedure for this which they should tell you about. Everyone should know where the book is kept or how the records are made. Anyone that has been hurt or has taken part in dealing with an occurrence should complete the details of what has happened. Typically this will require you to fill in:

* the date, time and place of the incident

* how it happened

* what was the cause

* how it was dealt with

* who was involved

* signature and date.

The details in the book have to be transferred onto an official HSE report form.

As far as is possible, the site, company or workplace will have set procedures in place for reporting hazards and accidents. These procedures will usually be found in the place where the accident book or records are stored. The location tends to be posted on the site notice board.

How hazards are created

Construction sites are busy places. There are constantly new stages in development. As each stage is begun a whole new set of potential hazards need to be considered.

At the same time, new workers will always be joining the site. It is mandatory for them to be given health and safety instruction during induction. But sometimes this is impossible due to pressure of work or availability of trainers.

Construction sites can become even more hazardous in times of extreme weather:

* Flooding – long periods of rain can cause trenches to fill with water, cellars to be flooded and smooth surfaces to become extremely wet and slippery.

* Wind – strong winds may prevent all work at height. Scaffolding may have become unstable, unsecured roofing materials may come loose, dry-stored materials such as sand and cement may have been blown across the site.

* Heat – this can change the behaviour of materials: setting quicker, failing to cure and melting. It can also seriously affect the health of the workforce through dehydration and heat exhaustion.

* Snow – this can add enormous weight to roofs and other structures and could cause collapse. Snow can also prevent access or block exits and can mean that simple and routine work becomes impossible due to frozen conditions.

Storing combustibles and chemicals

A combustible substance can be both flammable and explosive. There are some basic suggestions from the HSE about storing these:

* Ventilation – the area should be well ventilated to disperse any vapours that could trigger off an explosion.

* Ignition – an ignition is any spark or flame that could trigger off the vapours, so materials should be stored away from any area that uses electrical equipment or any tool that heats up.

* Containment – the materials should always be kept in proper containers with lids and there should be spillage trays to prevent any leak seeping into other parts of the site.

* Exchange – in many cases it can be possible to find an alternative material that is less dangerous. This option should be taken if possible.

* Separation – always keep flammable substances away from general work areas. If possible they should be partitioned off.

Combustible materials can include a large number of commonly used substances, such as cleaning agents, paints and adhesives.

HEALTH AND HYGIENE

Just as hazards can be a major problem on site, other less obvious problems relating to health and hygiene can also be an issue. It is both your responsibility and that of your employer to make sure that you stay healthy.

The employer will need to provide basic welfare facilities, no matter where you are working and these must have minimum standards.

KEY TERMS

Contamination

– this is when a substance has been polluted by some harmful substance or chemical.

Welfare facilities

Welfare facilities can include a wide range of different considerations, as can be seen in Table 1.7.

Facilities	Purpose and minimum standards
Toilets	If there is a lock on the door there is no need to have separate male and female toilets. There should be enough for the site workforce. If there is no flushing water on site they must be chemical toilets.
Washing facilities	There should be a wash basin large enough to be able to wash up to the elbow. There should be soap, hot and cold water and, if you are working with dangerous substances, then showers are needed.
Drinking water	Clean drinking water should be available; either directly connected to the mains or bottled water. Employers must ensure that there is no **contamination.**
Dry room	This can operate also as a store room, which needs to be secure so that workers can leave their belongings there and also use it as a place to dry out if they have been working in wet weather, in which case a heater needs to be provided.
Work break area	This is a shelter out of the wind and rain, with a kettle, a microwave, tables and chairs. It should also have heating.

Table 1.7 Welfare facilities in the workplace

CASE STUDY

South Tyneside Homes

South Tyneside Council's Housing Company

Staying safe on site

Johnny McErlane finished his apprenticeship at South Tyneside Homes a year ago.

'I've been working on sheltered accommodation for the last year, so there are a lot of vulnerable and elderly people around. All the things I learnt at college from doing the health and safety exams comes into practice really, like taking care when using extension leads, wearing high-vis and correct footwear. It's not just about your health and safety, but looking out for others as well.

On the shelters, you can get a health and safety inspector who just comes around randomly, so you have to always be ready. It just becomes a habit once it's been drilled into you. You're health and safety conscious all the time.

The shelters also have a fire alarm drill every second Monday, so you've got to know the procedure involved there. When it comes to the more specialised skills, such as mouth-to-mouth and CPR, you might have a designated first aider on site who will have their skills refreshed regularly. Having a full first aid certificate would be valuable if you're working in construction.

You cover quite a bit of the first aid skills in college and you really have to know them because you're not always working on large sites. For example, you might be on the repairs team, working in people's houses where you wouldn't have a first aider, so you've got to have the basic knowledge yourself, just in case. All our vans have a basic first aid kit that's kept fully stocked.

The company keeps our knowledge current with these "toolbox talks", which are like refresher courses. They give you any new information that needs to be passed on to all the trades. It's a good way of keeping everyone up to date.'

Noise

Ear defenders are the best precaution to protect the ears from loud noises on site. Ear defenders are either basic ear plugs or ear muffs, which can be seen in Fig 1.13 on page 32.

The long-term impact of noise depends on the intensity and duration of the noise. Basically, the louder and longer the noise exposure, the more damage is caused. There are ways of dealing with this:

* Remove the source of the noise.

* Move the equipment away from those not directly working with it.

* Put the source of the noise into a soundproof area or cover it with soundproof material.

* Ask a supervisor if they can move all other employees away from that part of the site until the noise stops.

Substances hazardous to health

COSHH Regulations (see page 3) identify a wide variety of substances and materials that must be labelled in different ways.

Controlling the use of these substances is always difficult. Ideally, their use should be eliminated (stopped) or they should be replaced with something less harmful. Failing this, they should only be used in controlled or restricted areas. If none of this is possible then they should only be used in controlled situations.

If a hazardous situation occurs at work, then you should:

* ensure the area is made safe

* inform the supervisor, site manager, safety officer or other nominated person.

You will also need to report any potential hazards or near misses.

Personal hygiene

Construction sites can be dirty places to work. Some jobs will expose you to dust, chemicals or substances that can make contact with your skin or may stain your work clothing. It is good practice to wear suitable PPE as a first line of defence as chemicals can penetrate your skin. Whenever you have finished a job you should always wash your hands. This is certainly true before eating lunch or travelling home. It can be good practice to have dedicated work clothing, which should be washed regularly.

Always ensure you wash your hands and face and scrub your nails. This will prevent dirt, chemicals and other substances from contaminating your food and your home.

Make sure that you regularly wash your work clothing and either repair it or replace it if it becomes too worn or stained.

Health risks

The construction industry uses a wide variety of substances that could harm your health. You will also be carrying out work that could be a health risk to you, and you should always be aware that certain activities could cause long-term damage or even kill you if things go wrong. Unfortunately not all health risks are immediately obvious. It is important to make sure that from time to time you have health checks, particularly if you have been using hazardous substances. Table 1.8 outlines some potential health risks in a typical construction site.

KEY TERMS

Dermatitis

– this is an inflammation of the skin. The skin will become red and sore, particularly if you scratch the area. A GP should be consulted.

Leptospirosis

– this is also known as Weil's disease. It is spread by touching soil or water contaminated with the urine of wild animals infected with the leptospira bacteria. Symptoms are usually flu-like but in extreme cases it can cause organ failure.

Health risk	Potential future problems
Dust	The most dangerous potential dust is, of course, asbestos, which **should only be handled by specialists under controlled conditions**. But even brick dust and other fine particles can cause eye injuries, problems with breathing and even cancer.
Chemicals	Inhaling or swallowing dangerous chemicals could cause immediate, long-term damage to lungs and other internal organs. Skin problems include burns or skin can become very inflamed and sore. This is known as **dermatitis**.
Bacteria	Contact with waste water or soil could lead to a bacterial infection. The germs in the water or dirt could cause infection which will require treatment if they enter the body. The most extreme version is **leptospirosis**.
Heavy objects	Lifting heavy, bulky or awkward objects can lead to permanent back injuries that could require surgery. Heavy objects can also damage the muscles in all areas of the body.
Noise	Failure to wear ear defenders when you are exposed to loud noises can permanently affect your hearing. This could lead to deafness in the future.
Vibrating tools	Using machines that vibrate can cause a condition known as hand/arm vibration syndrome (HAVS) or vibration white finger, which is caused by injury to nerves and blood vessels. You will feel tingling that could lead to permanent numbness in the fingers and hands, as well as muscle weakness.
Cuts	Any open wound, no matter how small, leaves your body exposed to potential infections. Cuts should always be cleaned and covered, preferably with a waterproof dressing. The blood loss from deep cuts could make you feel faint and weak, which may be dangerous if you are working at height or operating machinery.
Sunlight	Most construction work involves working outside. There is a temptation to take advantage of hot weather and get a tan. But long-term exposure to sunshine means risking skin cancer so you should cover up and apply sun cream.
Head injuries	You should seek medical attention after any bump to the head. Severe head injuries could cause epilepsy, hearing problems, brain damage or death.

Table 1.8 Health risks in construction

HANDLING AND STORING MATERIALS AND EQUIPMENT

On a busy construction site it is often tempting not to even think about the potential dangers of handling equipment and materials. If something needs to be moved or collected you will just pick it up without any thought. It is also tempting just to drop your tools and other equipment when you have finished with them to deal with later. But abandoned equipment and tools can cause hazards both for you and for other people.

Safe lifting

Lifting or handling heavy or bulky items is a major cause of injuries on construction sites. So whenever you are dealing with a heavy load, it is important to carry out a basic risk assessment.

The first thing you need to do is to think about the job to be done and ask:

* Do I need to lift it manually or is there another way of getting the object to where I need it?

Consider any mechanical methods of transporting loads or picking up materials. If there really is no alternative, then ask yourself:

1. Do I need to bend or twist?
2. Does the object need to be lifted or put down from high up?
3. Does the object need to be carried a long way?
4. Does the object need to be pushed or pulled for a long distance?
5. Is the object likely to shift around while it is being moved?

If the answer to any of these questions is 'yes', you may need to adjust the way the task is done to make it safer.

Think about the object itself. Ask:

1. Is it just heavy or is it also bulky and an awkward shape?
2. How easy is it to get a good hand-hold on the object?
3. Is the object a single item or are there parts that might move around and shift the weight?
4. Is the object hot or does it have sharp edges?

Again, if you have answered 'yes' to any of these questions, then you need to take steps to address these issues.

It is also important to think about the working environment and where the lifting and carrying is taking place. Ask yourself:

1. Are the floors stable?

2. Are the surfaces slippery?

3. Will a lack of space restrict my movement?

4. Are there any steps or slopes?

5. What is the lighting like?

Before lifting and moving an object, think about the following:

● Check that your pathway is clear to where the load needs to be taken.

● Look at the product data sheet and assess the weight. If you think the object is too heavy or difficult to move then ask someone to help you. Alternatively, you may need to use a mechanical lifting device.

When you are ready to lift, gently raise the load. Take care to ensure the correct posture – you should have a straight back, with your elbows tucked in, your knees bent and your feet slightly apart.

Once you have picked up the load, move slowly towards your destination. When you get there, make sure that you do not drop the load but carefully place it down.

DID YOU KNOW?

Although many people regard the weight limit for lifting and/or moving heavy or awkward objects to be 20 kg, the HSE does not recommend safe weights. There are many things that will affect the ability of an individual to lift and carry particular objects and the risk that this creates, so manual handling should be avoided altogether where possible.

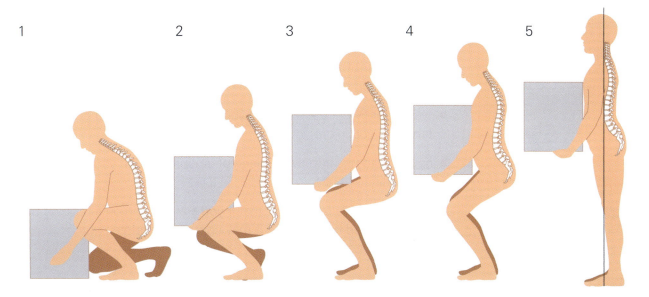

Figure 1.4 Take care to follow the correct procedure for lifting

Sack trolleys are useful for moving heavy and bulky items around. Gently slide the bottom of the sack trolley under the object and then raise the trolley to an angle of 45° before moving off. Make sure that the object is properly balanced and is not too big for the trolley.

Trailers and forklift trucks are often used on large construction sites, as are dump trucks. Never use these without proper training.

Figure 1.5 Pallet truck

Figure 1.6 Sack trolley

Site safety equipment

You should always read the construction site safety rules and when required wear your PPE. Simple things, such as wearing the right footwear for the right job, are important.

Safety equipment falls into two main categories:

* PPE – including hard hats, footwear, gloves, glasses and safety vests

* perimeter safety – this includes screens, netting and guards or clamps to prevent materials from falling or spreading.

Construction safety is also directed by signs, which will highlight potential hazards.

Safe handling of materials and equipment

All tools and equipment are potentially dangerous. It is up to you to make sure that they do not cause harm to yourself or others. You should always know how to use tools and equipment. This means either instruction from someone else who is experienced, or at least reading the manufacturer's instructions.

You should always make sure that you:

* use the right tool – don't be tempted to use a tool that is close to hand instead of the one that is right for the job

* wear your PPE – the one time you decide not to bother could be the time that you injure yourself

* never try to use a tool or a piece of equipment that you have not been trained to use.

You should always remember that if you are working on a building that was constructed before 2000 it may contain asbestos.

Correct storage

We have already seen that tools and equipment need to be treated with respect. Damaged tools and equipment are not only less effective at doing their job, they could also cause you to injure yourself.

Table 1.9 provides some pointers on how to store and handle different types of materials and equipment.

Materials and equipment	Safe storage and handling
Hand tools	Store hand tools with sharp edges either in a cover or a roll. They should be stored in bags or boxes. They should always be dried before putting them away as they will rust.
Power tools	Never carry them by the cable. Store them in their original carrying case. Always follow the manufacturer's instructions.
Wheelbarrows	Check the tyres and metal stays regularly. Always clean out after use and never overload.
Bricks and blocks	Never store more than two packs high. When cutting open a pack, be careful as the bricks could collapse.
Slabs and curbs	Store slabs flat on their edges on level ground, preferably with wood underneath to prevent damage. Store curbs the same way. To prevent weather damage, cover them with a sheet.
Tiles	Always cover them and protect them from damage as they are relatively fragile. Ideally store them in a hut or container.
Aggregates	Never store aggregates under trees as leaves will drop on them and contaminate them. Cover them with plastic sheets.
Plaster and plasterboard	Plaster needs to be kept dry, so even if stored inside you should take the precaution of putting the bags on pallets. To prevent moisture do not store against walls and do not pile higher than five bags. Plasterboard can be awkward to manage and move around. It also needs to be stored in a waterproof area. It should be stored flat and off the ground but should not be stored against walls as it may bend. Use a rotation system so that the materials are not stored in the same place for long periods.
Wood	Always keep wood in dry, well-ventilated conditions. If it needs to be stored outside it should be stored on bearers that may be on concrete. If wood gets wet and bends it is virtually useless. Always be careful when moving large cuts of wood or sheets of ply or MDF as they can easily become damaged.
Adhesives and paint	Always read the manufacturer's instructions. Ideally they should always be stored on clearly marked shelves. Make sure you rotate the stock using the older stock first. Always make sure that containers are tightly sealed. Storage areas must comply with fire regulations and display signs to advise of their contents.

Table 1.9 Safe storing and handling of materials and equipment

Waste control

The expectation within the building services industry is increasingly that working practices conserve energy and protect the environment. Everyone can play a part in this. For example, you can contribute by turning off hose pipes when you have finished using water, or not running electrical items when you don't need to.

Simple things, such as keeping construction sites neat and orderly, can go a long way to conserving energy and protecting the environment. A good way to remember this is Sort, Set, Shine, Standardise:

* Sort – sort and store items in your work area, eliminate clutter and manage deliveries.

* Set – everything should have its own place and be clearly marked and easy to access. In other words, be neat!

Figure 1.7 It's important to create as little waste as possible on the construction site

* Shine – clean your work area and you will be able to see potential problems far more easily.

* Standardise – by using standardised working practices you can keep organised, clean and safe.

Reducing waste is all about good working practice. By reducing wastage disposal, and recycling materials on site, you will benefit from savings on raw materials and lower transportation costs.

Planning ahead, and accurately measuring and cutting materials, means that you will be able to reduce wastage.

BASIC WORKING PLATFORMS AND ACCESS EQUIPMENT

Working at height should be eliminated or the work carried out using other methods where possible. However, there may be situations where you may need to work at height. These situations can include:

* roofing

* repair and maintenance above ground level

* working on high ceilings.

Any work at height must be carefully planned. Access equipment includes all types of ladder, scaffold and platform. You must always use a working platform that is safe. Sometimes a simple step ladder will be sufficient, but at other times you may have to use a tower scaffold.

Generally, ladders are fine for small, quick jobs of less than 30 minutes. However, for larger, longer jobs a more permanent piece of access equipment will be necessary.

Working platforms and access equipment: good practice and dangers of working at height

Table 1.10 outlines the common types of equipment used to allow you to work at heights, along with the basic safety checks necessary.

Equipment	Main features	Safety checks
Step ladder	Ideal for confined spaces. Four legs give stability	• Knee should remain below top of steps • Check hinges, cords or ropes • Position only to face work
Ladder	Ideal for basic access, short-term work. Made from aluminium, fibreglass or wood	• Check rungs, tie rods, repairs, and ropes and cords on stepladders • Ensure it is placed on firm, level ground • Angle should be no greater than 75° or 1 in 4
Mobile mini towers or scaffolds	These are usually aluminium and foldable, with lockable wheels	• Ensure the ground is even and the wheels are locked • Never move the platform while it has tools, equipment or people on it
Roof ladders and crawling boards	The roof ladder allows access while crawling boards provide a safe passage over tiles	• The ladder needs to be long enough and supported • Check boards are in good condition • Check the welds are intact • Ensure all clips function correctly
Mobile tower scaffolds	These larger versions of mini towers usually have edge protection	• Ensure the ground is even and the wheels are locked • Never move the platform while it has tools, equipment or people on it • Base width to height ratio should be no greater than 1:3
Fixed scaffolds and edge protection	Scaffolds fitted and sized to the specific job, with edge protection and guard rails	• There needs to be sufficient braces, guard rails and scaffold boards • The tubes should be level • There should be proper access using a ladder
Mobile elevated work platforms	Known as scissor lifts or cherry pickers	• Specialist training is required before use • Use guard rails and toe boards • Care needs to be taken to avoid overhead hazards such as cables

Table 1.10 Equipment for working at height and safety checks

You must be trained in the use of certain types of access equipment, like mobile scaffolds. Care needs to be taken when assembling and using access equipment. These are all examples of good practice:

* Step ladders should always rest firmly on the ground. Only use the top step if the ladder is part of a platform.

* Do not rest ladders against fragile surfaces, and always use both hands to climb. It is best if the ladder is steadied (footed) by someone at the foot of the ladder. Always maintain three points of contact – two feet and one hand.

* A roof ladder is positioned by turning it on its wheels and pushing it up the roof. It then hooks over the ridge tiles. Ensure that the access ladder to the roof is directly beside the roof ladder.

* A mobile scaffold is put together by slotting sections until the required height is reached. The working platform needs to have a suitable edge protection such as guard-rails and toe-boards. Always push from the bottom of the base and not from the top to move it, otherwise it may lean or topple over.

Figure 1.8 A tower scaffold

WORKING SAFELY WITH ELECTRICITY

It is essential whenever you work with electricity that you are competent and that you understand the common dangers. Electrical tools must be used in a safe manner on site. There are precautions that you can take to prevent possible injury, or even death.

Precautions

Whether you are using electrical tools or equipment on site, you should always remember the following:

* Use the right tool for the job.

* Use a transformer with equipment that runs on 110V.

* Keep the two voltages separate from each other. You should avoid using 230V where possible but, if you must, use a residual current device (RCD) if you have to use 230V.

* When using 110V, ensure that leads are yellow in colour.

* Check the plug is in good order.

* Confirm that the fuse is the correct rating for the equipment.

* Check the cable (including making sure that it does not present a tripping hazard).

* Find out where the mains switch is, in case you need to turn off the power in the event of an emergency.

* Never attempt to repair electrical equipment yourself.

* Disconnect from the mains power before making adjustments, such as changing a drill bit.

* Make sure that the electrical equipment has a sticker that displays a recent test date.

Visual inspection and testing is a three-stage process:

1. The user should check for potential danger signs, such as a frayed cable or cracked plug.

2. A formal visual inspection should then take place. If this is done correctly then most faults can be detected.

3. Combined inspections and **PAT** should take place at regular intervals by a competent person.

Watch out for the following causes of accidents – they would also fail a safety check:

KEY TERMS

PAT

– Portable Appliance Testing – regular testing is a health and safety requirement under the Electricity at Work Regulations (1989).

- damage to the power cable or plug
- taped joints on the cable
- wet or rusty tools and equipment
- weak external casing
- loose parts or screws
- signs of overheating
- the incorrect fuse
- lack of cord grip
- electrical wires attached to incorrect terminals
- bare wires.

When preparing to work on an electrical circuit, do not start until a permit to work has been issued by a supervisor or manager to a competent person.

Make sure the circuit is broken before you begin. A 'dead' circuit will not cause you, or anybody else, harm. These steps must be followed:

- Switch off – ensure the supply to the circuit is switched off by disconnecting the supply cables or using an isolating switch.
- Isolate – disconnect the power cables or use an isolating switch.
- Warn others – to avoid someone reconnecting the circuit, place warning signs at the isolation point.
- Lock off – this step physically prevents others from reconnecting the circuit.
- Testing – is carried out by electricians but you should be aware that it involves three parts:
  1. testing a voltmeter on a known good source (a live circuit) so you know it is working properly
  2. checking that the circuit to be worked on is dead
  3. rechecking your voltmeter on the known live source, to prove that it is still working properly.

It is important to make sure that the correct point of isolation is identified. Isolation can be next to a local isolation device, such as a plug or socket, or a circuit breaker or fuse.

The isolation should be locked off using a unique key or combination. This will prevent access to a main isolator until the work has been completed. Alternatively, the handle can be made detachable in the OFF position so that it can be physically removed once the circuit is switched off.

Dangers

You are likely to encounter a number of potential dangers when working with electricity on construction sites or in private houses. Table 1.11 outlines the most common dangers.

DID YOU KNOW?

All power tools should be checked by the user before use. A PAT programme of maintenance, inspection and testing is necessary. The frequency of inspection and testing will depend on the appliance. Equipment is usually used for a maximum of three months between tests.

Danger	Identifying the danger
Faulty electrical equipment	Visually inspect for signs of damage. Equipment should be double insulated or incorporate an earth cable.
Damaged or worn cables	Check for signs of wear or damage regularly. This includes checking power tools and any wiring in the property.
Trailing cables	Cables lying on the ground, or worse, stretched too far, can present a tripping hazard. They could also be cut or damaged easily.
Cables and pipe work	Always treat services you find as though they are live. This is very important as services can be mistaken for one another. You may have been trained to use a cable and pipe locator that finds cables and metal pipes.
Buried or hidden cables	Make sure you have plans. Alternatively, use a cable and pipe locator, mark the positions, look out for signs of service connection cables or pipes and hand-dig trial holes to confirm positions.
Inadequate over-current protection	Check circuit breakers and fuses are the correct size current rating for the circuit. A qualified electrician may have to identify and label these.

Table 1.11 Common dangers when working with electricity

Each year there are around 1,000 accidents at work involving electric shocks or burns from electricity. If you are working in a construction site you are part of a group that is most at risk. Electrical accidents happen when you are working close to equipment that you think is disconnected but which is, in fact, live.

Another major danger is when electrical equipment is either misused or is faulty. Electricity can cause fires and contact with the live parts can give you an electric shock or burn you.

Different voltages

The two most common voltages that are used in the UK are 230V and 110V:

* 230V: this is the standard domestic voltage. But on construction sites it is considered to be unsafe and therefore 110V is commonly used.

* 110V: these plugs are marked with a yellow casement and they have a different shaped plug. A transformer is required to convert 230V to 110V.

Some larger homes, as well as industrial and commercial buildings, may have 415V supplies. This is the same voltage that is found on overhead electricity cables. In most houses and other buildings the voltage from these cables is reduced to 230V. This is what most electrical equipment works from. Some larger machinery actually needs 415V.

In these buildings the 415V comes into the building and then can either be used directly or it is reduced so that normal 230V appliances can be used.

Colour coded cables

Normally you will come across three differently coloured wires: Live, Neutral and Earth. These have standard colours that comply with European safety standards and to ensure that they are easily identifiable. However, in some older buildings the colours are different.

Wire type	Modern colour	Older colour
Live	Brown	Red
Neutral	Blue	Black
Earth	Yellow and Green	Yellow and Green

Table 1.12 Colour coding of cables

Working with equipment with different electrical voltages

You should always check that the electrical equipment that you are going to use is suitable for the available electrical supply. The equipment's power requirements are shown on its rating plate. The voltage from the supply needs to match the voltage that is required by the equipment.

Storing electrical equipment

Electrical equipment should be stored in dry and secure conditions. Electrical equipment should never get wet but – if it does happen – it should be dried before storage. You should always clean and adjust the equipment before connecting it to the electricity supply.

PERSONAL PROTECTIVE EQUIPMENT (PPE)

Personal protective equipment, or PPE, is a general term that is used to describe a variety of different types of clothing and equipment that aim to help protect against injuries or accidents. Some PPE you will use on a daily basis and others you may use from time to time. The type of PPE you wear depends on what you are doing and where you are. For example, the practical exercises in this book were photographed at a college, which has rules and requirements for PPE that are different to those on large construction sites. Follow your tutor's or employer's instructions at all times.

Types of PPE

PPE literally covers from head to foot. Here are the main PPE types.

Figure 1.9 A hi-vis jacket

Figure 1.10 Safety glasses and goggles

Figure 1.11 Hand protection

Figure 1.12 Head protection

Figure 1.13 Hearing protection

Protective clothing

Clothing protection such as overalls:

* provides some protection from spills, dust and irritants
* can help protect you from minor cuts and abrasions
* reduces wear to work clothing underneath.

Sometimes you may need waterproof or chemical-resistant overalls.

High visibility (hi-vis) clothing stands out against any background or in any weather conditions. It is important to wear high visibility clothing on a construction site to ensure that people can see you easily. In addition, workers should always try to wear light-coloured clothing underneath, as it is easier to see.

You need to keep your high visibility and protective clothing clean and in good condition.

Employers need to make sure that employees understand the reasons for wearing high visibility clothing and the consequences of not doing so.

Eye protection

For many jobs, it is essential to wear goggles or safety glasses to prevent small objects, such as dust, wood or metal, from getting into the eyes. As goggles tend to steam up, particularly if they are being worn with a mask, safety glasses can often be a good alternative.

Hand protection

Wearing gloves will help to prevent damage or injury to the hands or fingers. For example, general purpose gloves can prevent cuts, and rubber gloves can prevent skin irritation and inflammation, such as contact dermatitis caused by handling hazardous substances. There are many different types of gloves available, including specialist gloves for working with chemicals.

Head protection

Hard hats or safety helmets are compulsory on building sites. They can protect you from falling objects or banging your head. They need to fit well and they should be regularly inspected and checked for cracks. Worn straps mean that the helmet should be replaced, as a blow to the head can be fatal. Hard hats bear a date of manufacture and should be replaced after about 3 years.

Hearing protection

Ear defenders, such as ear protectors or plugs, aim to prevent damage to your hearing or hearing loss when you are working with loud tools or are involved in a very noisy job.

Respiratory protection

Breathing in fibre, dust or some gases could damage the lungs. Dust is a very common danger, so a dust mask, face mask or respirator may be necessary.

Make sure you have the right mask for the job. It needs to fit properly otherwise it will not give you sufficient protection.

Foot protection

Foot protection is compulsory on site, particularly if you are undertaking heavy work. Footwear should include steel toecaps (or equivalent) to protect feet against dropped objects, midsole protection (usually a steel plate) to protect against puncture or penetration from things like nails on the floor and soles with good grip to help prevent slips on wet surfaces.

Figure 1.14 Respiratory protection

Legislation covering PPE

The most important piece of legislation is the Personal Protective Equipment at Work Regulations (1992). It covers all sorts of PPE and sets out your responsibilities and those of the employer. Linked to this are the Control of Substances Hazardous to Health (2002) and the Provision and Use of Work Equipment Regulations (1992 and 1998).

Storing and maintaining PPE

All forms of PPE will be less effective if they are not properly maintained. This may mean examining the PPE and either replacing or cleaning it, or if relevant testing or repairing it. PPE needs to be stored properly so that it is not damaged, contaminated or lost. Each type of PPE should have a CE mark. This shows that it has met the necessary safety requirements.

Importance of PPE

PPE needs to be suitable for its intended use and it needs to be used in the correct way. As a worker or an employee you need to:

* make sure you are trained to use PPE
* follow your employer's instructions when using the PPE and always wear it when you are told to do so
* look after the PPE and if there is a problem with it report it.

Your employer will:

* know the risks that the PPE will either reduce or avoid
* know how the PPE should be maintained
* know its limitations.

Consequences of not using PPE

The consequences of not using PPE can be immediate or long-term. Immediate problems are more obvious, as you may injure yourself. The longer-term consequences could be ill health in the future. If your employer has provided PPE, you have a legal responsibility to wear it.

FIRE AND EMERGENCY PROCEDURES

If there is a fire or an emergency, it is vital that you raise the alarm quickly. You should leave the building or site and then head for the **assembly point.**

When there is an emergency a general alarm should sound. If you are working on a larger and more complex construction site, evacuation may begin by evacuating the area closest to the emergency. Areas will then be evacuated one-by-one to avoid congestion of the escape routes.

Figure 1.15 Assembly point sign

Three elements essential to creating a fire

Three ingredients are needed to make something combust (burn):

* oxygen   * heat   * fuel.

The fuel can be anything which burns, such as wood, paper or flammable liquids or gases, and oxygen is in the air around us, so all that is needed is sufficient heat to start a fire.

The fire triangle represents these three elements visually. By removing one of the three elements the fire can be prevented or extinguished.

Figure 1.16 The fire triangle

How fire is spread

Fire can easily move from one area to another by finding more fuel. You need to consider this when you are storing or using materials on site, and be aware that untidiness can be a fire risk. For example, if there are wood shavings on the ground the fire can move across them, burning up the shavings.

Heat can also transfer from one source of fuel to another. If a piece of wood is on fire and is against or close to another piece of wood, that too will catch fire and the fire will have spread.

On site, fires are classified according to the type of material that is on fire. This will determine the type of fire-fighting equipment you will need to use. The five different types of fire are shown in Table 1.13.

Class of fire	Fuel or material on fire
A	Wood, paper and textiles
B	Petrol, oil and other flammable liquids
C	LPG, propane and other flammable gases
D	Metals and metal powder
E	Electrical equipment

Table 1.13 Different classes of fire

There is also F, cooking oil, but this is less likely to be found on site, except in a kitchen.

Taking action if you discover a fire and fire evacuation procedures

During induction, you will have been shown what to do in the event of a fire and told about assembly points. These are marked by signs and somewhere on the site there will be a map showing their location.

If you discover a fire you should:

* sound the alarm

* not attempt to fight the fire unless you have had fire marshal training

* otherwise stop work, do not collect your belongings, do not run, and do not re-enter the site until the all clear has been given.

Different types of fire extinguishers

Extinguishers can be effective when tackling small localised fires. However, you must use the correct type of extinguisher. For example, putting water on an oil fire could make it explode. For this reason, you should not attempt to use a fire extinguisher unless you have had proper training.

When using an extinguisher it is important to remember the following safety points:

* Only use an extinguisher at the early stages of a fire, when it is small.

* The instructions for use appear on the extinguisher.

* If you do choose to fight the fire because it is small enough, and you are sure you know what is burning, position yourself between the fire and the exit, so that if it doesn't work you can still get out.

Type of fire risk	Fire class Symbol	White label Water	Cream label Foam	Black label Carbon dioxide	Blue label Dry powder	Yellow label Wet chemical
A – Solid (e.g. wood or paper)	A	✓	✓	✗	✓	✓
B – Liquid (e.g. petrol)	B	✗	✓	✓	✓	✗
C – Gas (e.g. propane)	C	✗	✗	✓	✓	✗
D – Metal (e.g. aluminium)	D METAL	✗	✗	✗	✓	✗
E – Electrical (i.e. any electrical equipment)	E	✗	✗	✓	✓	✗
F – Cooking oil (e.g. a chip pan)	F	✗	✗	✗	✗	✓

Table 1.14 Types of fire extinguishers

There are some differences you should be aware of when using different types of extinguisher:

* CO_2 extinguishers – do not touch the nozzle; simply operate by holding the handle. This is because the nozzle gets extremely cold when ejecting the CO_2, as does the canister. Fires put out with a CO_2 extinguisher may reignite, and you will need to ventilate the room after use.

* Powder extinguishers – these can be used on lots of kinds of fire, but can seriously reduce visibility by throwing powder into the air as well as on the fire.

SIGNS AND SAFETY NOTICES

In a well-organised working environment safety signs will warn you of potential dangers and tell you what to do to stay safe. They are used to warn you of hazards. Their purpose is to prevent accidents. Some will tell you what to do (or not to do) in particular parts of the site and some will show you where things are, such as the location of a first aid box or a fire exit.

Types of signs and safety notices

There are five basic types of safety sign, as well as signs that are a combination of two or more of these types. These are shown in Table 1.15.

Type of safety sign	What it tells you	What it looks like	Example
Prohibition sign	Tells you what you must *not* do	Usually round, in red and white	Do not use ladder
Hazard sign	Warns you about hazards	Triangular, in yellow and black	Caution Slippery floor
Mandatory sign	Tells you what you *must* do	Round, usually blue and white	Masks must be worn in this area
Safe condition or information sign	Gives important information, e.g. about where to find fire exits, assembly points or first aid kit, or about safe working practices	Green and white	First aid
Firefighting sign	Gives information about extinguishers, hydrants, hoses and fire alarm call points, etc.	Red with white lettering	Fire alarm call point
Combination sign	These have two or more of the elements of the other types of sign, e.g. hazard, prohibition and mandatory		DANGER Isolate before removing cover

Table 1.15 Different types of safety signs

TEST YOURSELF

1. Which of the following requires you to tell the HSE about any injuries or diseases?

   a. HASAWA

   b. COSHH

   c. RIDDOR

   d. PUWER

2. What is a prohibition notice?

   a. An instruction from the HSE to stop all work until a problem is dealt with

   b. A manufacturer's announcement to stop all work using faulty equipment

   c. A site contractor's decision not to use particular materials

   d. A local authority banning the use of a particular type of brick

3. Which of the following is considered a major injury?

   a. Bruising on the knee

   b. Cut

   c. Concussion

   d. Exposure to fumes

4. If there is an accident on a site who is likely to be the first to respond?

   a. First aider

   b. Police

   c. Paramedics

   d. HSE

5. Which of the following is a summary of risk assessments and is used for high risk activities?

   a. Site notice board

   b. Hazard book

   c. Monitoring statement

   d. Method statement

6. Some substances are combustible. Which of the following are examples of combustible materials?

   a. Adhesives

   b. Paints

   c. Cleaning agents

   d. All of these

7. What is dermatitis?

   a. Inflammation of the skin

   b. Inflammation of the ear

   c. Inflammation of the eye

   d. Inflammation of the nose

8. Screens, netting and guards on a site are all examples of which of the following?

   a. PPE

   b. Signs

   c. Perimeter safety

   d. Electrical equipment

9. Which of the following are also known as scissor lifts or cherry pickers?

   a. Bench saws

   b. Hand-held power tools

   c. Cement additives

   d. Mobile elevated work platforms

10. In older properties the neutral electricity wire is which colour?

    a. Black

    b. Red

    c. Blue

    d. Brown

Chapter 2

Unit CSA–L3Core07

ANALYSING TECHNICAL INFORMATION, QUANTITIES AND COMMUNICATION WITH OTHERS

LEARNING OUTCOMES

LO1: Know how to produce different types of drawings and information in the construction industry

LO2: Know how to estimate quantities and price work for contracts

LO3: Know how to ensure good working practices

INTRODUCTION

The aims of this chapter are to:

* help you to interpret information

* help you to estimate quantities

* help you to organise the building process and communicate the design work to colleagues and others.

PRODUCING DIFFERENT TYPES OF DRAWING AND INFORMATION

Accurate construction requires the creation of accurate drawings and matching supporting information. Supporting information can be found in a variety of different types of documents. These include:

* drawings and plans

* programmes of work

* procedures

* specifications

* policies

* schedules

* manufacturers' technical information

* organisational documentation

* training and development records

* risk and method statements

* Construction (Design and Management) (CDM) Regulations

* Building Regulations.

Each different type of construction plan has a definite look and purpose. Typical construction drawings focus in on floor plans or elevation.

Construction drawings are drawn to scale and need to be accurate, so that relative sizes are correct. The scale will be stated on the drawing, to avoid inaccuracies.

Working alongside these construction drawings are matched and linked specifications and schedules. It is these that outline all of the materials and tasks required to complete specific jobs.

In order to understand construction drawings you not only need to understand their purpose and what they are showing, but also a range of hatchings and symbols that act as shortcuts on the documents.

Electronic and traditional drawing methods

Construction drawings are only part of a long process in the design of buildings. In fact the construction drawings are the final stage or final version of these drawings. The design process begins with a basic concept, which is followed by outline drawings. By the end of the design stage working drawings, technical specifications and contract drawings have been completed.

The project is put out to tender. This is a process that involves companies bidding for the job based on the information that they have been given.

There are further changes just before the construction phase gets under way. The chosen construction company may have noted issues with the design, which means that the drawings may have to be amended. It is also at this stage that the construction company will begin the process of pricing up each phase of the job.

Electronic drawing methods

Many construction drawings are based on a system known as computer aided design (CAD). CAD basically produces two-dimensional electronic drawings using the similar lines, hatches and text that can be seen in traditional paper drawings.

Each different CAD drawing is created independently, so each design change has to be followed up on other CAD drawings.

Increasingly, however, a new electronic system is being used. It is known as building information modelling (BIM). This creates drawings in 3D. The buildings are virtually modelled from real construction elements, such as walls, windows and roofs. The big advantages are:

* it allows architects to design buildings in a similar way to the way in which the building will actually be built

* a central virtual building model stores all the data, so any changes to this are applied to individual drawings

* better coordinated designs can be created meaning that construction should be more straightforward.

Systems such as BIM provide 3D models, which can be viewed from any angle or perspective. It also includes:

* scheduling information

* labour required

* estimated costs

* a detailed breakdown of the construction phases.

Figure 2.1 BIM generated model

Traditional drawing methods

The development of the computer, laptops and hand-held tablets such as iPads is gradually making manual drafting of construction drawings obsolete. The majority of drawings are now created using CAD or BIM software.

Traditionally, drawings were limited to the available paper size and what would be convenient to transport.

As each of the traditional construction drawings were hand drawn, there was always a danger that the information on one drawing would not match the information on another. The only way to check that both were accurate was to cross-reference every detail.

One advantage is that paper plans are easier to carry around site, as computers can be broken or stolen. However, damaging or losing a paper plan can cause delays while it is replaced.

Types of supporting information

Drawings and plans

Drawings are an important part of construction work. You will need to understand how they provide you with the information you need to carry out the work. The drawings show what the building will look like and how it will be constructed. This means that there are several different drawings of the building from different viewpoints. In practice most of the drawings are shown on the same sheet.

Block plans

Block plans show the construction site and the surrounding area. Normally block plans are at a ratio of 1:1250 and 1:2500. This means that 1 millimetre on a block plan is equal to 1,250 mm (12.5 m) or 2,500 mm (25 m) or on the ground.

Figure 2.2 Block plan

Site plan

Often location drawings are also known as block plans or site plans. The site plan drawing shows what is planned for the site. It is often an important drawing because it has been created in order to get approval for the project from planning committees or funding sources. In most cases the site plan is actually an architectural plan, showing the basic arrangement of buildings and any landscaping.

The site plan will usually show:

* directional orientation (i.e. the north point)

* location and size of the building or buildings

* existing structures

* clear measurements.

General location

Location drawings show the site or building in relation to its surroundings. It will therefore show details such as boundaries, other buildings and roads. It will also contain other vital information, including:

* access

* drainage

* sewers

* the north point.

As with all plan drawings, the scale will be shown and the drawing will be given a title. It will be given a job or project number to help identify it easily, as well as an address, the date of the drawing and the name of the client. A version number will also be on the drawing with an amendment date if there have been any changes. You'll need to make sure you have the latest drawing.

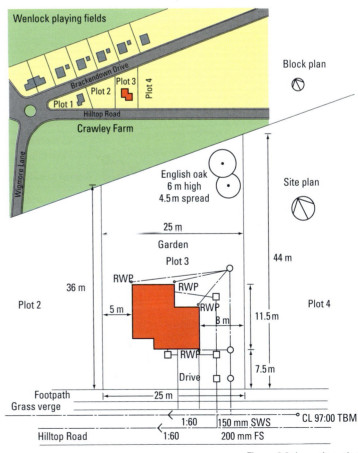

Figure 2.3 Location plan

Normally location drawings are either 1:200 or 1:500 (that is, 1 mm of the drawing represents 200 mm (2 m) or 500 mm (5 m) on the ground).

Assembly

These are detailed drawings that illustrate the different elements and components of the construction. They tend to be 1:5, 1:10 or 1:20 (1 mm of the drawing represents 5, 10 or 20 mm on the ground). This larger scale allows more detail to be shown, to ensure accurate construction.

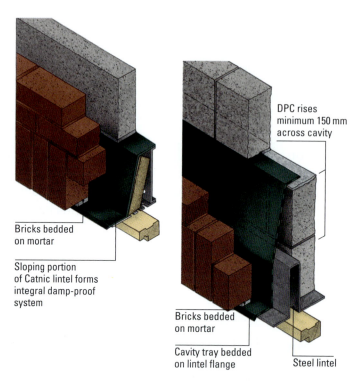

DPC rises minimum 150 mm across cavity

Bricks bedded on mortar

Sloping portion of Catnic lintel forms integral damp-proof system

Bricks bedded on mortar

Cavity tray bedded on lintel flange

Steel lintel

Figure 2.4 Assembly drawing

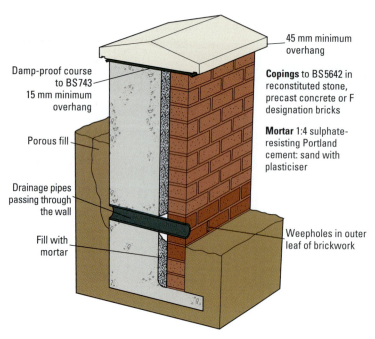

45 mm minimum overhang

Damp-proof course to BS743 15 mm minimum overhang

Copings to BS5642 in reconstituted stone, precast concrete or F designation bricks

Porous fill

Mortar 1:4 sulphate-resisting Portland cement: sand with plasticiser

Drainage pipes passing through the wall

Fill with mortar

Weepholes in outer leaf of brickwork

Sectional

These drawings aim to provide:

* vertical and horizontal measurements and details

* constructional details.

They can be used to show the height of ground levels, damp-proof courses, foundations and other aspects of the construction.

Figure 2.5 Section drawing of an earth retaining wall

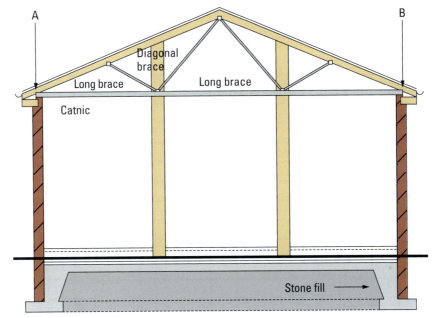

A

B

Diagonal brace

Long brace   Long brace

Catnic

Floor – slope to front 100mm concrete on 1200 gauge polythene, blinding and hardcore in layers

Stone fill →

Figure 2.6 Section drawing of a garage

Details

These drawings show how a component needs to be manufactured. They can be shown in various scales, but mainly 1:10, 1:5 and 1:1 (the same size as the actual component if it is small).

Programmes of work

Programmes of work show the actual sequence of any work activities on a construction project. Part of the work programme plan is to show target times. They are usually shown in the form of a Gantt chart (a special type of bar chart), as can be seen in Fig 2.8.

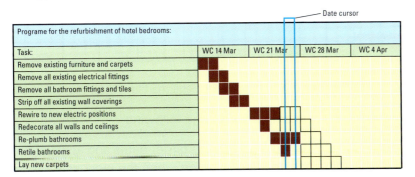

Figure 2.8 Single line contract plan Gantt chart

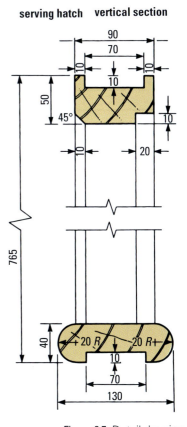

serving hatch   vertical section

Figure 2.7 Detail drawing

In this figure:

* on the left-hand side all of the tasks are listed – note this is in logical order

* on the right the blocks show the target start and end date for each of the individual tasks

* the timescale can be days, weeks or months.

Far more complex forms of work programmes can also be created. Fig 2.9 shows the planning for the construction of a house.

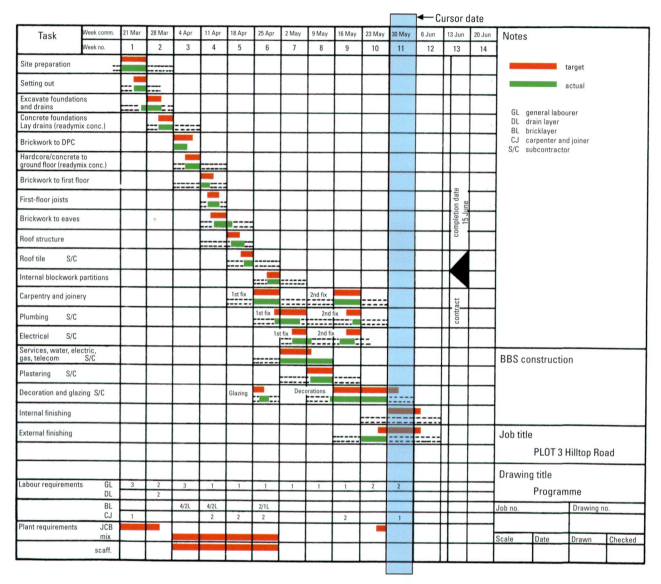

Figure 2.9 Gantt chart for the construction of a house

This more complex example shows the following:

● There are two lines – they show the target dates and actual dates. The actual dates are shaded, showing when the work actually began and how long it took.

● If this bar chart is kept up to date an accurate picture of progress and estimated completion time can be seen.

Procedures

When you work for a construction company they will have a series of procedures they will expect you to follow. A good example is the emergency procedure. This will explain precisely what is required in the case of an emergency on site and who will have responsibility to carry out particular duties. Procedures are there to show you the right way of doing something.

Another good example of a procedure is the procurement or buying procedure. This will outline:

● who is authorised to buy what, and how much individuals are allowed to spend

● any forms or documents that have to be completed when buying.

Specifications

In addition to drawings it is usually necessary to have documents known as specifications. These provide much more information, as can be seen in Fig 2.10.

The specifications give you a precise description. They will include:

● the address and description of the site

● on-site services (e.g. water and electricity)

● materials description, outlining the size, finish, quality and tolerances

● specific requirements, such as the individual that will authorise or approve work carried out

● any restrictions on site, such as working hours.

Policies

Policies are sets of principles or a programme of actions. The following are two good examples:

● environmental policy – how the business goes about protecting the environment

● safety policy – how the business deals with health and safety matters and who is responsible for monitoring and maintaining it.

You will normally find both policies and procedures in site rules. These are usually explained to each new employee when they first join the company. Sometimes there may be additional site rules, depending on the job and the location of the work.

Schedules

Schedules are cross-referenced to drawings that have been prepared by an architect. They will show specific design information. Usually they are prepared for jobs that will crop up regularly on site, such as:

● working on windows, doors, floors, walls or ceilings

● working on drainage, lintels or sanitary ware.

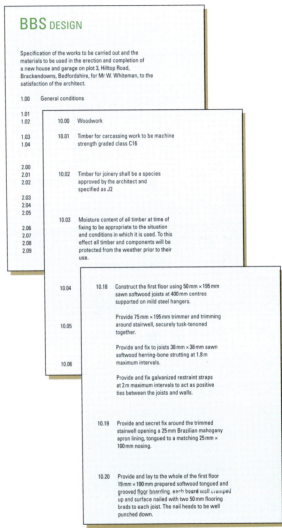

Figure 2.10 Extracts from a typical specification

A schedule can be seen in Fig 2.11.

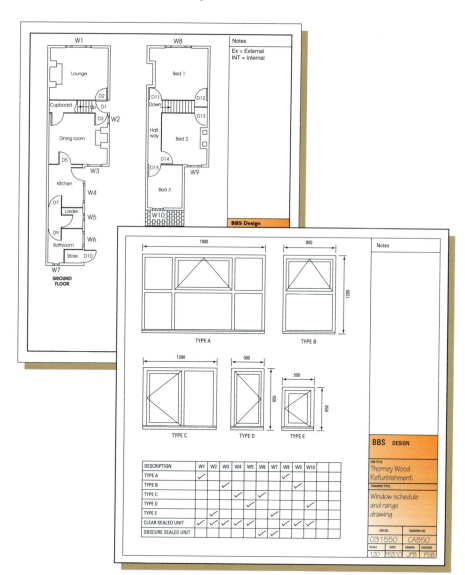

Figure 2.11 Typical windows schedule, range drawing and floor plans

The schedule is very useful for a number of reasons:

* working out the quantities of materials needed

* ordering materials and components and then checking them against deliveries

* locating where specific materials will be used.

Manufacturers' technical information

Almost everything that is bought to be used on site will come with a variety of information. The basic technical information provided will show what the equipment or material is intended to be used for, how it should be stored and any particular requirements it may have, such as handling or maintenance.

Technical information from the manufacturer can come from a variety of different sources:

* printed or downloadable data sheets

* printed or downloadable user instructions

* manufacturers' catalogues or brochures

* manufacturers' websites.

Organisational documentation

The potential list of organisational documentation and paperwork is massive. Examples are outlined in the following table.

Document	Purpose
Timesheet	Record of hours that you have worked and the jobs that you have carried out. They are used to help work out your wages and the total cost of the job.
Day worksheet	These detail work that has been carried out without providing an estimate beforehand. They usually include repairs or extra work and alterations.
Variation order	These are provided by the architect and given to the builder, showing any alterations, additions or omissions to the original job.
Confirmation notice	Provided by the architect to confirm any verbal instructions.
Daily report or site diary	Include things that might affect the project like detailed weather conditions, late deliveries or site visitors.
Orders and requisitions	These are order forms, requesting the delivery of materials.
Delivery notes	These are provided by the supplier of materials as a list of all materials being delivered. These need to be checked against materials actually delivered.
Delivery record	These are lists of all materials that have been delivered on site.
Memorandum	These are used for internal communications and are usually brief.
Letters	These are used for external communications, usually to customers or suppliers.
Fax	Even though email is commonly used, the industry still likes faxes, because they provide an exact copy of an original document.

Table 2.1

Training and development records

Training and development is an important part of any job, as it ensures that employees have all the skills and knowledge that they need to do their work. Most medium to large employers will have training policies that set out how they intend to do this.

To make sure that they are on track and to keep records they will have a range of different documents. These will record all the training that an employee has undertaken.

Training can take place in a number of different ways:

* induction

* toolbox talks

* in-house training

* specialist training

* training or education leading to formal qualifications.

Details required for floor plans

The floor plans shows the arrangement of the building, rather like a map. It is a cut through of the building, which shows openings, walls and other features usually at around 1 m above floor level.

The floor plan also includes elements of the building that can be seen below the 1 m level, such as the floor or part of the stairs. The drawing will show elements above the 1 m level as dotted lines. The floor plan is a vertical orthographic projection onto a horizontal plane. In effect the horizontal plane cuts through the building.

The floor plan will detail the following:

* Vertical and horizontal sections – these show the building cut along an axis to show the interior structure.

* Datum levels – these are taken from a nearby and convenient datum point. They show the building's levels in relation to the datum point.

* Wall constructions – this is revealed through the section or cross sections shown in the diagram. It details the wall construction methods and materials.

* Material codes – these will contain notes and links to specific materials and may also note particular parts of the Building Regulations that these construction materials comply with.

* Depth and height dimensions – these are drawn between the walls to show the room sizes and wall lengths. They are noted as width × depth.

* Schedules – these note repeated design information, such as types of door, windows and other features.

* Specifications – these outline the type, size and quality of materials, methods of fixing and quality of work and finish expected.

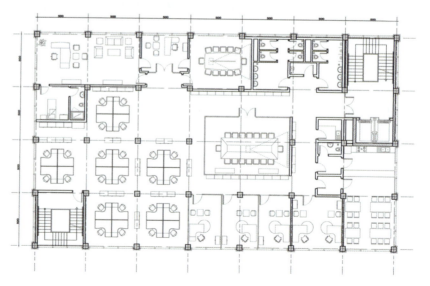

Figure 2.12 Example of a traditional floor plan

Details required for elevations

The details required for elevations in construction drawings are the same as those required for floor plans. An elevation is the view of the building as seen from one side. It can be used to show what the exterior of the building will look like. The elevation is labelled in relation to the compass direction. The elevation is a horizontal orthographic projection of the building onto a vertical plane. Usually the vertical plane is parallel to one side of the building (orthographic drawing).

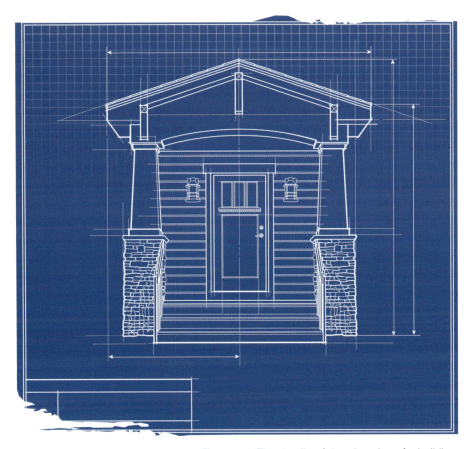

Figure 2.13 The details of the elevation of a building

Linking schedules to drawings

The schedule of work and the drawings create a single set of information. These documents need to be clear and comprehensive.

Before construction gets under way the specification schedule is the most important set of documents. It is used by the construction company to price up the job, work out how to tackle it, and then put in a bid for the work.

The construction company can look at each task in detail and see what materials are needed. This, along with all the construction information documents, will help them to make an estimate as to how long the task will take to complete and to what standard it should be completed.

During the construction period the most important documents are the drawings. Each piece of work is linked to those drawings and a schedule of work is set up. This might incorporate a Gantt chart or critical path analysis, showing expected dates and duration of on-site and off-site activities. This might need a good deal of cross-referencing. Obviously you cannot fit windows until the relevant cavity wall has been built and the opening formed.

It is important that the drawings and the specification schedules are closely linked. Reference numbers and headings that are on the drawings need to appear with exactly the same numbers and words on the schedule. This will avoid any confusion. It should be possible to look at the drawings, find a reference number or heading and then look through the schedule to find the details of that particular task. It also allows the drawings to be slightly clearer, as they won't need to have detailed information on them that can be found in the schedule.

Reasons for different projections in construction drawings

Designers will use a range of drawings in order to get across their requirements. Each is a 2D image. They show what the building will look like, along with the components or layout.

Orthographic projections

Orthographic projections are used to show the different elevations or views of an object. Each of the views is at right angles to the face.

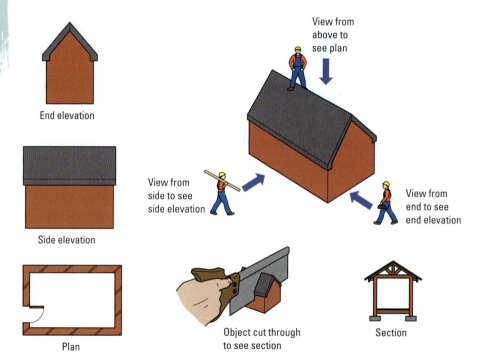

Figure 2.14 Plans, elevations and sections

Orthographic projection can be seen either as a first angle European projection or a third angle American projection. The following table shows the difference between these two views and there are examples in Figs 2.16 and 2.17, which relate to the shape shown in Fig 2.15.

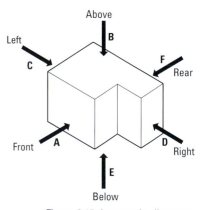

Figure 2.15 Isometric diagram showing the various views that can be portrayed in orthographic projection

Projection	Description
First angle	Everything is drawn in relation to the front view. The view from above is drawn below and the view from below is drawn from above. The view from the left is to the right and the right to the left. So all views, in effect, are reversed.
Third angle	This is often referred to as being an American projection. Everything again is in relation to the front elevation. The views from above and below are drawn in their correct position. Anything on the left is drawn to the left and the right to the right.

Table 2.2

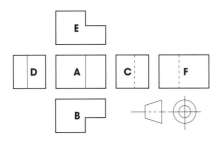

Figure 2.16 First angle projection

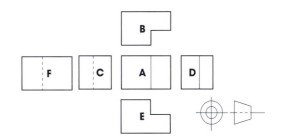

Figure 2.17 Third angle projection

Pictorial projections

Pictorial projections show objects in a 3D form. There are different ways of showing the view by varying the angles of the base line and the scale of any side projections. The most common is isometric. Vertical lines are drawn vertically, and horizontal lines are drawn at an angle of 30° to the horizontal. All of the other measurements are drawn to the same scale. This type of pictorial projection can be seen in Fig 2.18.

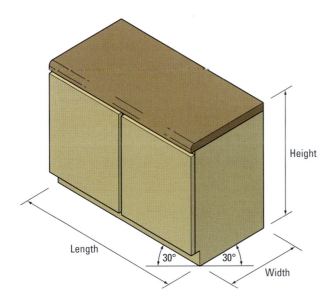

Figure 2.18 Isometric projection

There are four other different types of pictorial projection. These are not used as commonly as isometric projections.

Pictorial projection	Description
Planometric	Vertical lines are drawn vertically and horizontal lines on the front elevation of the object are drawn at 30°. The horizontal lines on the side elevation are drawn at 60° to horizontal.
Axonometric	The horizontal lines on all elevations are drawn at 45° to the horizontal. Otherwise the look is very similar to planometric.
Oblique	All of the vertical lines are drawn vertically. The horizontal lines on the front elevation are drawn horizontally but all the other horizontal lines are drawn at 45° to the horizontal.
Perspective	Horizontal lines are drawn so that they disappear into an imaginary horizon, known as a vanishing point. A one-point perspective drawing has all the sides disappearing to one vanishing point. An angular perspective, or two point perspective, has the elevations disappearing to two vanishing points.

Table 2.3

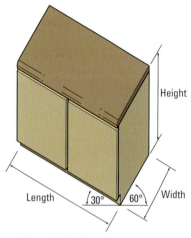

Figure 2.19 Planometric projection

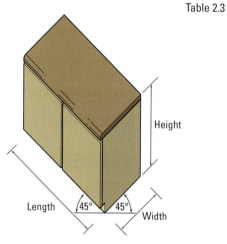

Figure 2.20 Axonometric projection

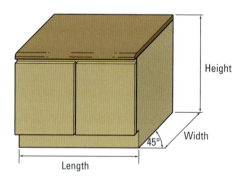

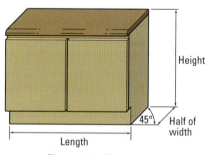

Figure 2.21 Oblique projection

VP = viewpoint

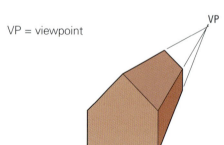

Figure 2.22 Parallel (one point) perspective projection

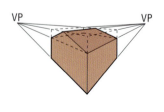

Figure 2.23 Angular (two point) perspective projection

Hatchings and symbols

Different materials and components are shown using symbols and hatchings. Abbreviations are also used. This makes the working drawings far less cluttered and easier to read.

Examples of symbols and abbreviations can be seen in Figs 2.24 and 2.25.

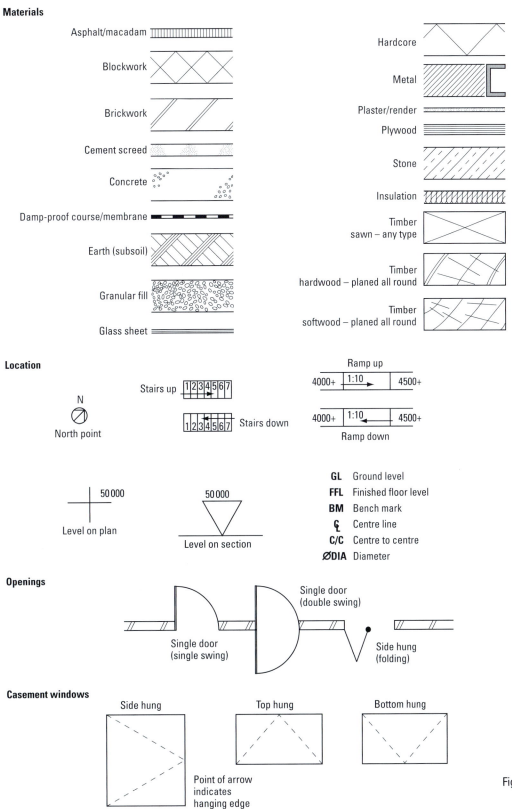

Figure 2.24 Symbols used on drawings

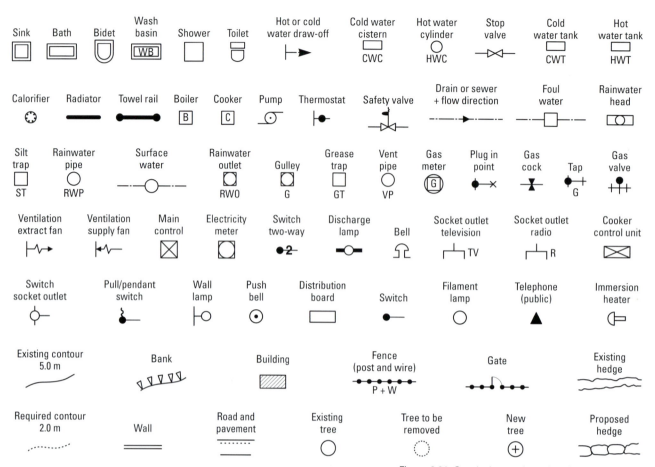

Figure 2.24 Symbols used on drawings *continued*

Aggregate	agg	BS tee	BST	Foundation	fdn	Polyvinyl acetate	PVA
Air brick	AB	Building	bldg	Fresh air inlet	FAI	Polyvinylchloride	PVC
Aluminium	al	Cast iron	CI	Glazed pipe	GP	Rainwater head	RWH
Asbestos	abs	Cement	ct	Granolithic	grano	Rainwater pipe	RWP
Asbestos cement	absct	Cleaning eye	CE	Hardcore	hc	Reinforced concrete	RC
Asphalt	asph	Column	col	Hardboard	hdbd	Rodding eye	RE
Bitumen	bit	Concrete	conc	Hardwood	hwd	Foul water sewer	FWS
Boarding	bdg	Copper	Copp cu	Inspection chamber	IC	Surface water sewer	SWS
Brickwork	bwk	Cupboard	cpd	Insulation	insul	Softwood	swd
BS* Beam	BSB	Damp-proof course	DPC	Invert	inv	Tongued and grooved	T&G
BS Universal beam	BSUB	Damp-proof membrane	DPM	Joist	jst	Unglazed pipe	UGP
BS Channel	BSC	Discharge pipe	DP	Mild steel	MS	Vent pipe	VP
BS equal angle	BSEA	Drawing	dwg	Pitch fibre	PF	Wrought iron	WI
BS unequal angle	BSUA	Expanding metal lathing	EML	Plasterboard	pbd		

Figure 2.25 Abbreviations commonly used on drawings

ESTIMATING QUANTITIES AND PRICING WORK FOR CONTRACTS

Working out the quantity and cost of resources that are needed to do a particular job can be difficult. In most cases you or the company you work for will be asked to provide a price for the work. It is generally accepted that there are three ways of doing this:

* Estimate – which is an approximate price, though estimation is a skill based on many factors.

* Quotation – which is a fixed price.

* Tender – which is a competitive quotation against other companies for a prescribed amount of work to a certain standard.

As we will see a little later in this section, these three ways of costing are very different and each of them has its own issues.

Resource requirements

As you become more experienced you will be able to estimate the amount of materials that will be needed on particular construction projects though this depends on the size and complexity of the job. This is also true of working out the best place to buy materials and how much the labour costs will be to get the job finished.

In order to work out how much a job will cost, you will need to know some basic information:

* What type of contract is agreed?

* What materials will be used?

* What are the costs of the materials?

Much of this information can be gained from the drawings, specification and other construction information for the proposed building

To help work out the price of a job, many businesses use the *UK Building Blackbook,* which provides a construction cost guide. It breaks down all types of work and shows an average cost for each of them.

Computerised estimating packages are available, which will give a comprehensive detailed estimate that looks very professional. This will also help to estimate quantities and timescales.

Measurement

The standard unit for measurement is based on the metre (m). There are 100 centimetres (cm) and 1,000 millimetres (mm) in a metre. It is important to remember that drawings and plans have different scales, so these need to be converted to work out quantities of materials.

The most basic thing to work out is length, from which you can calculate perimeter, area and then volume, capacity, mass and weight, as can be seen in the following table.

Measurement	Explanation
Length	This is the distance from one end to the other. For most jobs metres will be sufficient, although for smaller work such as brick length or lengths of screws, millimetres are used.
Perimeter	This is the distance around a shape, such as the size of a room or a garden. It will help you estimate the length of a wall, for example. You just need to measure each side and then add them together.
Area	You can work out the area of a room, for example, by measuring its length and its width. Then you multiply the width by the length to give the number of square metres (m²).
Volume and capacity	Volume shows how much space is taken up by an object, such as a room. Again this is simply worked out by multiplying the width of the room by its length and then by its height. This gives you the number of cubic metres (m³).

Capacity works in exactly the same way but instead of showing the figure as cubic metres you show it as litres. This is ideal if you are trying to work out the capacity of the water tank or a garden pond. |
| Mass or weight | Mass is measured usually in kilograms or in grams. Mass is the actual weight of a particular object, such as a brick. |

Table 2.4

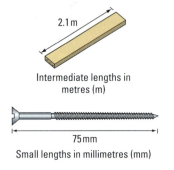

Figure 2.26 Length in metres and millimetres

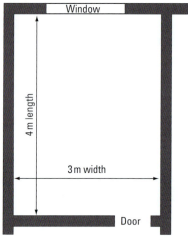

Figure 2.27 Measuring area and perimeter

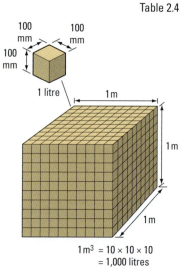

Figure 2.28 Relationship between volume and capacity

Formulae

These can appear to be complicated, but using formulae is essential for working out quantities of materials. Each formula is related to different shapes. In construction you will often have to work out quantities of materials needed for odd shaped areas.

Area

To work out the area of a triangular shape, you use the following formula:

$$\text{Area (A)} = \text{Base (B)} \times \text{Height (H)} \div 2$$

So if a triangle has a base of 4.5 and a height of 3.5 the calculation is:

$$4.5 \times 3.5 \div 2$$

Or $4.5 \times 3.5 = 15.75 \div 2 = 7.875\,m^2$

Height

If you want to work out the height of a triangle you switch the formulae around. To give us height = 2 × Area ÷ Base

Perimeter

To work out the perimeter of a rectangle use the formula:

$$\text{Perimeter} = 2 \times (\text{Length} + \text{Width})$$

It is important to remember this because you need to count the length and the width twice to ensure you have calculated the total distance around the object.

Circles

To work out the circumference or perimeter of a circle you use the formula:

$$\text{Circumference} = \pi \text{ (pi)} \times \text{diameter}$$

π (pi) is always the same for all circles and is 3.142.

Diameter is the length of the widest part.

If you know the circumference and need to work out the diameter of the circle the formula is:

$$\text{Diameter} = \text{circumference} \div \pi \text{ (pi)}$$

For example if a circle has a circumference of 15.39 m then to work out the diameter:

$$15.39 \div 3.142 = 4.89\,m$$

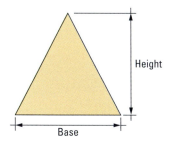

Figure 2.29 Triangle

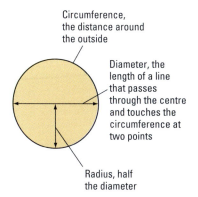

Circumference, the distance around the outside

Diameter, the length of a line that passes through the centre and touches the circumference at two points

Radius, half the diameter

Figure 2.30 Parts of a circle

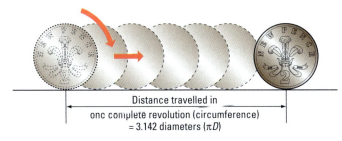

Distance travelled in one complete revolution (circumference) = 3.142 diameters (πD)

Figure 2.31 Relationship between circumference and diameter

Complex areas

Land, for example, is rarely square or rectangular. It is made up of odd shapes. Never be overwhelmed by complex areas, as all you need to do is to break them down into regular shapes.

By accurately measuring the perimeter you can then break down the shape into a series of triangles or rectangles. All you need to do then is to work out the area of each of the shapes within the overall shape and then add them together.

Shape		Area equals	Perimeter equals
Square 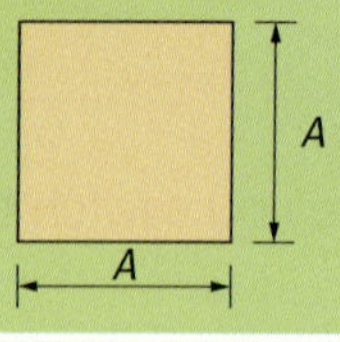		AA (or A multiplied by A)	4A (or A multiplied by 4)
Rectangle 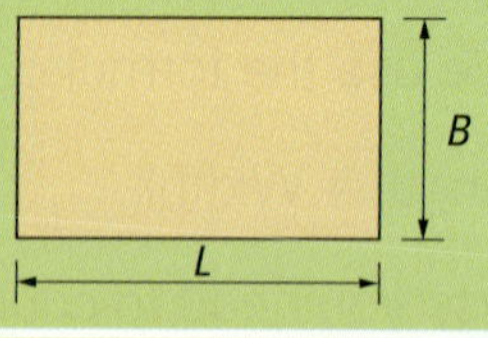		LB (or L multiplied by B)	2(L+B) (or L plus B multiplied by 2)
Trapezium 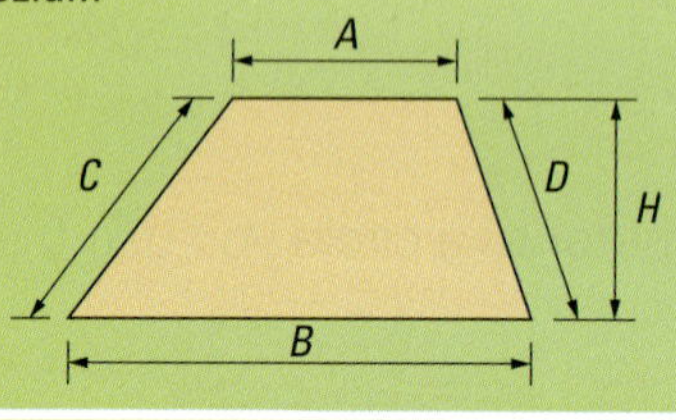		$\dfrac{(A + B)H}{2}$ (or A plus B multiplied by H and then divided by 2)	A+B+C+D
Triangle 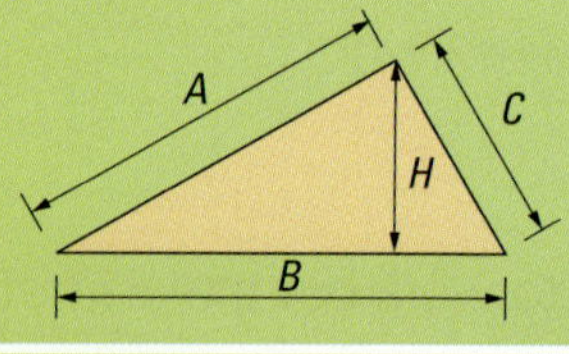		$\dfrac{BH}{2}$ (or B multiplied by H and then divided by 2)	A+B+C
Circle 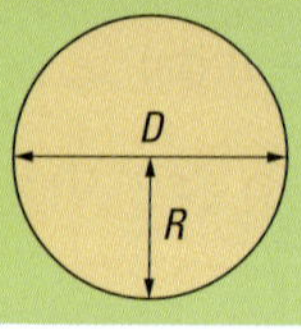		πR^2 (or R multiplied by itself and then multiplied by pi (3.142))	πD or $2\pi R$

Figure 2.32 Table of shapes and formulae

Volume

Sometimes it is necessary to work out the volume of an object, such as a cylinder or the amount of concrete needed. All that needs to be done is to work out the base area and then multiply that by the height.

For a concrete area, if a 1.2 m square needs 3 m of height then the calculation is:

$$1.2 \times 1.2 \times 3 = 4.32\,\text{m}^3$$

To work out the volume of a cylinder you need to know the base area × the height. The formula is:

$$\pi r^2 \times H$$

So if a cylinder has a radius (r) of 0.8 and a height of 3.5 m then the calculation is:

$$3.142 \times 0.8 \times 0.8 \times 3.5 = 7.038\,\text{m}^3$$

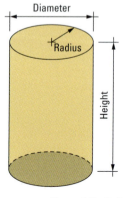

Figure 2.33 Cylinder

Pythagoras

Pythagoras' theorem is used to work out the length of the sides of right angled triangles. The theory states that:

In all right angled triangles the square of the longest side is equal to the sum of the squares of the other two sides (that is, the length of a side multiplied by itself).

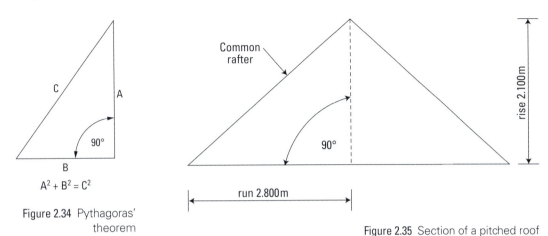

$$A^2 + B^2 = C^2$$

Figure 2.34 Pythagoras' theorem

Figure 2.35 Section of a pitched roof

See Chapter 7 for more about geometry, trigonometry and Pythagoras.

Measuring materials

Using simple measurements and formulae can help you work out the amount of materials you will need. This is all summarised in the following table.

Material	Measurement
Timber	Can be sold by the cubic metre. To work out the length of material divide the cross section area of one section by the total cross section area of the material.
Flooring	To work out the amount of flooring for a particular area multiply the width of the floor by the length of the floor.
Stud walling	Measure the distance that the stud partition will cover then divide that distance by a specified spacing. This will give you the number of spaces between each stud.
Rafters and floor joists	Measure the distance between the adjacent walls then take into account that the first and last joist or rafter will be 50 mm away from the wall. Measure the total distance and then divide it by the specified spacing.
Fascias, barges and soffits	Measure the length and then add a little extra to take into account any necessary cutting and jointing.
Skirting	You need to work out the perimeter of the room and then subtract any doorways or other openings. This technique can be used to work out the necessary length of dado, picture rails and coving.
Bricks and mortar	Half-brick walls use 60 bricks per metre squared and one-brick walls use double that amount. You should add 5 per cent to take into account any cutting or damage. For mortar assume that you will need 1 kg for each brick.

Table 2.5

How to cost materials

Once you have found out the quantity of materials necessary you will need to find out the price of those materials. It is then simply a case of multiplying those prices by the amount of materials actually needed to find out approximately how much they will cost in total.

Materials and purchasing systems

Many builders and companies will have preferred suppliers of materials. Many of them will already have negotiated discounts based on their likely spending with that supplier over the course of a year. The supplier will be geared up to supply them at an agreed price.

In other cases builders may shop around to find the best price for the materials that match the specification. It is not always the case that the lowest price is necessarily the best. All materials need to be of a sufficient quality. The other key consideration is whether the materials are immediately available for delivery.

It is vital that suppliers are reliable and that they have sufficient materials in stock. Delays in deliveries can cause major setbacks on site. It is not always possible to warn suppliers that materials will be needed, but a well-run site should be able to anticipate the materials that are needed and put in the orders within good time.

Large quantities may be delivered direct from the manufacturer straight to site. This is preferable when dealing with items where consistency, for example of colour, is required.

Labour rates and costs

The cost of labour for particular jobs is based on the hourly charge-out rate for that individual or group of individuals multiplied by the time it would take to complete the job.

Labour rates can depend on the:

* expertise of the construction worker

* size of the business they work for

* part of the country in which the work is being carried out

* complexity of the work.

According to the International Construction Costs Survey 2012, the following were average costs per hour:

* Group 1 tradespeople – plumbers, electricians etc.: £30

* Group 2 tradespeople – carpenters, bricklayers etc.: £30

* Group 3 tradespeople – tillers, carpet layers and plasterers: £30

* general labourers: £18

* site supervisors: £46.

REED TIP

A great career path can start with an apprenticeship. 80 per cent of the staff at South Tyneside Homes started off as apprentices. Some have worked their way up to job roles such as team leaders, managers and heads of departments.

Quotes, estimated prices and tenders

As we have already seen, estimates, quotes and tenders are very different. We need to look at these in slightly more detail, as can be seen in the following table.

Type of costing	Explanation
Estimate	This needs to be a realistic and accurate calculation based on all the information available as to how much a job will cost. An estimate is not binding and the client needs to understand that the final cost might be more.
Quote	This is a fixed price based on a fixed specification. The final price may be different if the fixed specification changes; for example if the customer asks for additional work then the price will be higher.
Tender	This is a competitive process. The customer advertises the fact that they want a job done and invites tenders. The customer will specify the specifications and schedules and may even provide the drawings. The companies tendering then prepare their own documents and submit their price based on the information the customer has given them. All tenders are submitted to the customer by a particular date and are sealed. The customer then opens all tenders on a given date and awards the contract to the company of their choice. This process is particularly common among public sector customers such as local authorities.

Table 2.6

Inaccurate estimates

Larger companies will have an estimating team. Smaller businesses will have someone who has the job of being an estimator. Whenever they are pricing a job, whether it is a quote, an estimate or a tender, they will have to work out the costs of all materials, labour and other costs. They will also have to include a **mark-up.**

It is vital that all estimating is accurate. Everything needs to be measured and checked. All calculations need to be double-checked.

It can be disastrous if these figures are wrong because:

* if the figure is too high then the client is likely to reject the estimate and look elsewhere as some competitors could be cheaper

* if the figure is too low then the job may not provide the business with sufficient profit and it will be a struggle to make any money out of the job.

KEY TERMS

Mark-up

– a builder or building business, just like any other business, needs to make a profit. Mark-up is the difference between the total cost of the job and the price that the customer is asked to pay for the work.

DID YOU KNOW?

Many businesses fail as a result of not working out their costs properly. They may have plenty of work but they are making very little money.

CASE STUDY

South Tyneside *Homes*

South Tyneside Council's

Bringing all your skills together to do a good job

Marcus Chadwick, a bricklayer at Laing O'Rourke, talks about maths and English skills.

'Obviously you need your maths, especially being a bricklayer. From the dimensions on your drawings, you have to be able to work out how many materials you'll need – how many bricks, how many blocks, how much sand and cement you need to order. Eventually it ends up being rote learning, like the way you learn your times-tables. With a bit of practice, you'll be able to work out straight away, "Right, I need x number of blocks, I need 1000 bricks there, I need a ton of sand, therefore I need seven bags of cement." Though there are still times when I get the calculator out!

If you get it wrong and miscalculate it can delay the progression of the building, or your section. I'm the foreman and if I set out a wall in the wrong position then there's only one person to blame. So you check, then double-check – it's like the old saying, "Measure twice, cut once".

Number skills really are important; you can't just say "Well, I'm a bricklayer and I'm just going to work with my hands". But that all comes with time; I wasn't that good at maths when I left school, though it was probably a case of just being lazy. When you come into an environment where you need to start using it to earn the money, then you'll start to get it straight away.

Your communication skills are definitely important too. You need to know how to speak to people. I always find that your lads appreciate you more if you ask them to do something as opposed to tell them. That was your old 1970s mentality where you used to have your screaming foreman – "Get this done, get that done!" – it doesn't work like that anymore. You have to know how to speak to people, to communicate.

All the lads that work for me, they're my extended family. My boss knows that too and that's why I've been with him for seven years now – he takes me everywhere because he knows I've got a good relationship with all my workforce.

You'll also have the odd occasion where the client will come around to visit the site and you've got to be able to put yourself across, using good diction. That also goes for when you're ordering materials – because you deal with different regions across the country, you've got to be clearly spoken so they understand you, so things don't get messed up in translation.'

Purchasing or hiring plant and equipment

Normally, if a piece of plant or equipment is going to be used on a regular basis then it is purchased by the company. By maximising the use of any plant or equipment, the business will save on the costs of repeatedly hiring and the transport of the item to and from the site.

It also does not make sense to leave plant and equipment on a site if it is no longer being used. It needs to be moved to a new site where it can be used.

Many smaller construction companies have no alternative other than to hire. This is because they cannot afford to have an enormous amount of money tied up in the plant or equipment, whether this comes from earned profits or from a loan or finance agreement. Loans and finance agreements have to be paid back over a period of time.

The decision as to whether to purchase or to hire is influenced by a number of factors:

* The working lives of the plant or equipment – how long will it last? This will usually depend on how much it is used and how well maintained it is.

* The use of the plant or equipment – is the company going to get good use out of it if they buy it? If they are hiring it then it should only be hired for the time it is actually needed. There is no point in having the plant or equipment on site and paying for its hire if it is not being used.

* Loss of value – just like buying a brand new car, the value of new plant or equipment takes an enormous drop the moment you take delivery of it. Even if it is hardly used it is considered second-hand and is not worth anything like its price when it was new. The biggest falls in value are in the first few years that the company owns it. It then reaches a value that it will sit at for some years until it is considered junk or scrap.

* Obsolescence – what might seem today to be the most advanced and technologically superior piece of plant or equipment may not be so tomorrow. Newer versions will come onto the market and may be more efficient or cost-effective. It is probable that the plant or equipment will be obsolete, or outdated, before it ends its useful working life.

* Cost of replacement – investing in plant or equipment today means that at some point in the future they will have to be replaced. The business will have to take account of this and arrange to have the necessary funds available for replacement in the future.

* Maintenance costs – if the construction business owns the plant or equipment they will have to pay for any routine maintenance, repairs and of course operators. Hired equipment, such as diggers or cranes, is the responsibility of the hiring company. They pay for all the maintenance and although they charge for the operator, the operator is on their wage bill.

* Insurance and licences – owning plant or equipment often means additional insurance payments and the company may also have to obtain licences that allow them to use that type of equipment in a particular area. Hired plant and equipment is already insured and should have the relevant licences.

* Financial costs – if the decision is to buy rather than hire, the money that would have otherwise been sitting in a bank account, earning interest, has been spent. If the company had to borrow the money to buy the plant or equipment then interest charges are payable on loans and finance agreements.

PRACTICAL TIP

Many construction companies that know they are going to be working on a project for a long period of time will actually buy plant and equipment for that contract. Once the contract has been completed they will sell on the plant and equipment.

Planning the sequence of materials and labour requirements

One of the most important jobs when organising work that will need to be carried out on site is to calculate when, where and how much materials and labour will be needed at any one time. This is organised in a number of different ways. The following headings cover the main documents or processes that are involved.

Bill of quantities

BILL OF QUANTITIES				
Contract		DWG No.		
DESCRIPTION	QUANTITY	UNIT	RATE	AMOUNT

Figure 2.36 Bill of quantities form

This is used by building contractors when they quote for work on larger projects. It is usually prepared by a quantity surveyor. Fig 2.36 shows you what a bill of quantities looks like.

The form is completed using information from the working drawings, specification and schedule (this is called the take off). It describes each particular job and how many times that job needs to be carried out. It sets the number of units of material or labour, the rate at which they are charged and the total amount.

Programmes of work

A programme of work is also an important document, as it looks at the length of time and the sequence of jobs that will be needed to complete the construction. It has three main sections:

* A master programme that shows the start and finish dates. It shows the duration, sequence and any relationships between jobs across the whole contract.

* A stage programme – this is the next level down and it covers particular stages of the contract. A good example would be the foundation work or the process of making the building weather-tight. Alternatively it might look at a period of up to two months' worth of work in detail.

* A weekly programme – there will be several of these, which aim to predict where and when work will take place across the whole of the site. These are very important as they need to be compared against actual progress. The normal process is to review and update these weekly programmes and then update the stage and master programmes if delays have been encountered.

Stock systems and lead times

One of the greatest sources of delays in construction is not having the right materials and equipment available when it is needed. This means that someone has to work out not only what is needed and how many, but when. It is a balancing act because there are dangers in having the stock on site too early. If all the materials needed for a construction job arrived in the first week then this would cause problems and it is unlikely that there would be anywhere to store them. Materials need to be ordered to ensure that they are on site just before they are needed.

One of the problems is lead times. There is no guarantee that the supplier will have sufficient stock available when you need it. They need to be warned that you will need a certain amount of material at a certain time in advance. This will allow them to either manufacture the stock or get it from their supplier. Specialist materials have longer lead times. These may have to be specially manufactured, or perhaps imported from abroad. All of this takes time.

Once the quantities of materials have been calculated and the sequence of work decided, comparing that to the duration of the project and the schedule, it should be possible to predict when materials will be needed. You will need to liaise with your suppliers as soon as possible to find out the lead times they need to get the materials delivered to the site. This might mean that you will have to order materials out of sequence to the work schedule because some materials need longer lead times than others.

Planning and scheduling using charts

To plan the sequence of materials and labour requirements it is often a good idea to put the information in a format that can be easily read and understood. This is why many companies use charts, graphs and other types of illustrated diagram.

The most common is probably the Gantt chart. This is a series of horizontal bars. Each different task or operation involved in a project is shown on the left-hand side of the chart. Along the top are days, weeks or months. The planner marks the start day, week or month and the projected end day, week or month with a horizontal bar. It shows when tasks start and when they end. It will also be useful in showing when labour will be needed. It will also show which jobs have to be completed before another job can begin. An example of a Gantt chart can be seen in Fig 2.37.

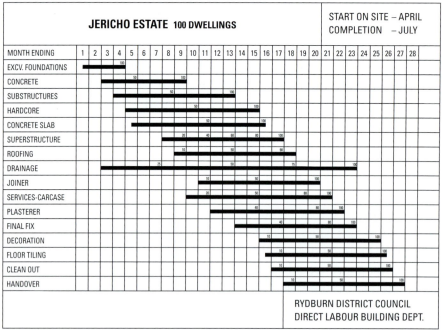

Typical programme for rate of completion on a housing development contract

Figure 2.37 Gantt chart

There are other types of bar chart that can be used to plan and monitor work on the construction site. It is important to remember that each chart relates to the plan of work:

* A single bar – this focuses in on a sequence of tasks and the bar is filled in to show progress.

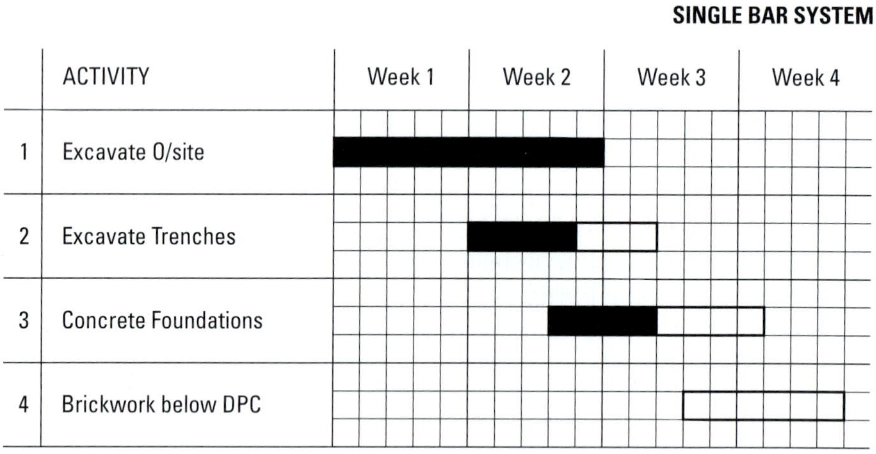

SINGLE BAR SYSTEM

	ACTIVITY	Week 1	Week 2	Week 3	Week 4
1	Excavate O/site				
2	Excavate Trenches				
3	Concrete Foundations				
4	Brickwork below DPC				

Figure 2.38 Single bar chart

- A two-bar – this tracks the amount of work that has been carried out against the planned amount of work that should have been carried out. In other words it shows the percentage of work that has been completed. It is there to alert the site manager that work may be falling behind and extra resources are needed.

TWO BAR SYSTEM

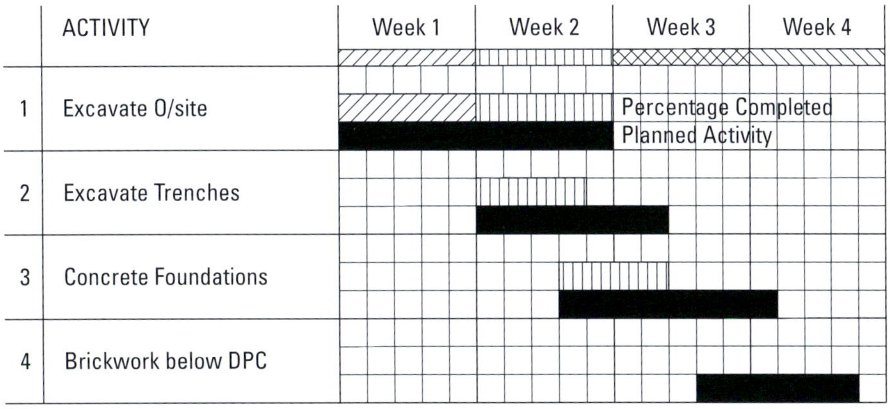

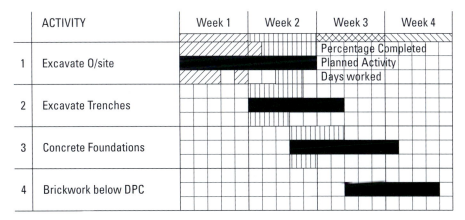

Figure 2.39 Two-bar system

- A three-bar – this shows the planned duration of the activity, the actual days that have seen work being done on that activity and the percentage of the activity that has been completed. This gives a snapshot view of how work has progressed over the course of a period of time.

THREE BAR SYSTEM

Figure 2.40 Three-bar system

Calculating hours required

We have already seen that different types of construction workers attract different hourly rates of pay. The simple solution in order to work out the cost to complete particular work is to look at the programmes of work and the estimated time required to complete it. The next stage is to estimate how many workers will be needed to carry out that activity and then multiply that by the estimated labour cost per hour.

Added costs

When a construction company estimates the costs of work they have to incorporate a number of other different costs. A summary of these can be seen in the following table.

Added cost	Explanation
National Insurance contributions	Companies employing workers have to pay National Insurance contributions to the government for each employee.
Value Added Tax (VAT)	For businesses that are registered for VAT they have to charge a sales tax on any services that they provide. They collect this money on behalf of the government. The VAT is added to the final cost of the work.
Pay As You Earn (PAYE)	PAYE, or income tax, has to be paid on the income of all workers straight to the government.
Travel expenses	This is particularly relevant if workers on site have to travel a considerable distance in order to do their work. This can reasonably be passed on to the client.
Profit and loss	Many businesses will make the mistake of trying to estimate the costs of their work in the knowledge that they are competing with other businesses, trimming their estimates if they can. As we will see when we look at profitability, a business does need to make money from contracts otherwise it will not have sufficient funds to continue to operate.
Suppliers' terms and conditions	Suppliers will often have set payment terms, such as 30, 60 or 90 days. The construction company needs to have sufficient funds to pay their suppliers on the due date stated on the invoice. If they fail to do this then they could run into difficulties as the supplier may decide not to give them any more credit until outstanding invoices have been paid.
Wastage	It is rarely possible to buy materials and components that are an exact fit. Blocks and bricks, for example, will have a percentage damaged or unusable on each pallet. Materials can also be mis-cut, damaged or otherwise wasted on site. It is sensible for the construction company to factor in a wastage rate of at least 5 per cent. This is often required to allow for cutting and fitting when using stock lengths.
Penalty clauses	Many projects are time sensitive and need to be completed by specified dates. The contracts will state whether there are any penalties to be made if critical dates are missed. Penalty clauses are rather like fines that the construction company pays if they fail to meet deadlines.

Table 2.7

Total estimated prices

As we have seen, the total estimated price needs to incorporate all of the added costs. But there are two other issues that should not be forgotten:

* The cost of any plant and equipment hire – the length of time that these are necessary will have to be calculated together with an additional period in case of delays.

* Contingencies – it is not always possible to determine exact prices, especially in groundworks, and, in any case, not all construction jobs will run smoothly. It is therefore sensible to set aside funds should additional work be needed. This may be particularly true if additional work needs to be done to secure the foundations or if workers and equipment are on site yet it is not possible to work due to poor weather conditions.

Profitability

Setting an initial price for a business's services is, perhaps, one of the most difficult tasks. It needs to take account of the costs that are incurred by the business. It also needs to consider the prices charged by key competitors, as the optimum price that a business wishes to charge may not be possible if competitors are charging considerably lower prices.

It is difficult for a new business to set prices, because its services may not be well known. Its costs may be comparatively higher because it will not have the advantages of providing services on a large scale. Equally, it cannot charge high prices because neither the company nor its services are established in the market.

The process of calculating revenue is a relatively simple task. Sales revenue or income is equal to the services sold, multiplied by the average selling price. In other words, all a business needs to know is how many services have been sold, or might sell, and the price they will charge.

We have already seen that a business incurs costs that must be paid. Clearly these have a direct impact on the profitability of a business.

We can already see that there is a direct relationship between costs and profit. Costs cut into the revenue generated by the business and reduce its overall profitability.

Gross profit is the difference between a company's revenue and its costs. Businesses will also calculate their operating profit. The operating profit is the business's gross profit minus its **overheads.**

A business may also calculate its pre-tax profits, which are its profits before it pays its taxes. It may have one-off costs, such as the replacement of a piece of plant or equipment. These costs are deducted from the operating profit to give the pre-tax profit.

KEY TERMS

Overheads

– these are expenses that need to be paid by the business regardless of how much work they have on at any one time, such as the rent of builders' yard.

The business will then pay tax on the remainder of its profit. This will leave them their net profit. This is the amount of money that they have actually made over the course of a year or on a particular job.

Profits are an important measure of the success of a business. Like other businesses construction companies do borrow money, but profit is the source of around 60 per cent of the funds that businesses use to help them grow.

Businesses can look for ways to gradually increase their profit. They can look at each type of job they do and work out the most efficient way of doing it. This might mean looking for a particular mix of employees, or buying or hiring particular plant and equipment that will speed up the work.

GOOD WORKING PRACTICES

Like any business, construction relies on a number of factors to make sure that everything runs smoothly and that a company's reputation is maintained.

There needs to be a good working relationship between those who work for the construction company, and other individuals or companies that they regularly deal with, such as the local authority, and professionals such as architects and clients.

This is achieved by making sure that these individuals and organisations continue to have trust and confidence in the company. Any promises or guarantees that are made must be kept.

In the normal course of events communication needs to be clear and straightforward. When there are problems accurate and honest communication can often deal with many of them. It can set aside the possibility of misunderstandings.

Good working relationships

Each construction job will require the services of a team of professionals. They will need to be able to work and communicate effectively with one another. Each has different roles and responsibilities.

Although you probably won't be working with exactly the same people all the time you are on site, you will be working with on-site colleagues every day. These may be people doing the same job as you, as well as people with other roles and responsibilities who you need to work with to ensure that the project runs smoothly.

Working with unskilled operatives

It's important to remember that everyone will have different levels of skill and experience. You, as a skilled or trade operative, are qualified in your trade, or working towards your qualification. Some people will be less experienced than you; for example, unskilled operatives (manual workers) are entry level operatives without any formal training. They may, however, be experienced on sites and will take instructions from the supervisor or site manager. You should be patient with colleagues who are less experienced or skilled than you – after all, everyone has to learn. However, if you see them carrying out unsafe practices, you should tell your supervisor or charge-hand straight away.

Working with skilled employees

You'll also work with people who are more experienced than you. It's a good idea to watch how they work and learn from their example. Show them respect and don't expect to know as much as they do if they have been working for much longer. However, if you see them ignoring safety rules, don't copy them; speak to your supervisor.

Working with professional technicians

You might also work with professional technicians, such as civil engineers or architectural technicians. They will have extensive knowledge in their field but may not know as much as you about bricklaying or carpentry. For your relationship to run smoothly, you should respect each other's knowledge, share your thoughts on any issues, and listen to what each other has to say.

Working with supervisors

Supervisors organise the day-to-day running of the site or a team. Charge-hands supervise a specific trade, such as bricklayers or carpenters. They will be your immediate boss, and you must listen to their instructions and obey any rules they set out. These rules enable the site to be run smoothly and safely so it is in your interest to do what your supervisor says.

Working with the site manager

The site manager or site agent runs the construction site, makes plans to avoid problems and meet deadlines, and ensures all processes are carried out safely. They communicate directly with the client. They are ultimately responsible for everything that goes on at the construction site. Even if you don't communicate with them directly, you should follow the guidance and rules that they have put in place. It's in your interest to do your bit to keep the site safe and efficient.

Working with other professionals

You may also need to work with or communicate with other professionals. For example, a clerk of works is employed by the architect on behalf of a client. They oversee the construction work and ensure that it represents the interests of the client and follows agreed specifications and designs. A contracts manager agrees prices and delivery dates. These professionals will expect you to do the job that has been specified and to draw their attention to anything that will change the plans they are responsible for.

Hierarchical charts

A hierarchy describes the different levels of responsibility, authority and power in a business or organisation. The larger the business the more levels of management it will have. The higher up the management structure the more responsibility each person will have.

Decisions are made at the top and instructions are passed down the hierarchy. It is best to imagine most organisations as like a pyramid. The directors or owners of the business are at the top of that pyramid. A site manager may be part-way down and at the bottom of the pyramid are all the workers who are on site.

Trust and confidence

The trust of colleagues develops as a result of showing that the company is reliable, cooperative and committed to the success or goals of the colleague or client. Trust does not happen automatically but has to be earned through actions. An important part of this is building a positive relationship with colleagues. Over time, trust will develop into confidence. Colleagues will have confidence in the company being able to deliver their promises.

The reputation of a construction company is very important. Criticisms of the company will always do far more damage than the positive benefits of a successfully completed contract. This shows how important it is to get things right the first time and every time.

In an industry where there is so much competition, trust and confidence can mean the difference between getting the contract and being rejected before your bid has even been considered.

As the construction company becomes established they will build up a network of colleagues. If these colleagues have trust and confidence in the business they will recommend the business to others.

Ultimately, earning trust and confidence relies on the business being able to solve any problems with the minimum of fuss and delay. Fair solutions need to be identified. These solutions should not be against the interests of anyone involved.

Accurate communication

Effective communication in all types of work is essential. It needs to be clear and to the point, as well as accurate. Above all it needs to be a two-way process. This means that any communication that you have with anyone must be understood.

In construction work it is essential to keep to deadlines and follow strict instructions and specifications. Failing to communicate will always cause confusion, extra cost and delays. In an industry such as this it is unacceptable and very easy to avoid. Negative communication or poor communication can damage the confidence that others have in you to do your job.

It is important to have a good working relationship with colleagues at work. An important part of this is to communicate in a clear way with them. This helps everyone understand what is going on, what decisions have been made. It also means being clear. Most communication with colleagues will be verbal (spoken). Good communication results in:

* cutting out mistakes and stoppages (saving money)

* avoiding delays

* making sure that the job is done right the first time and every time.

The more complex a contract, the more likely it is that changes and alterations will be needed. The longer the contract runs for, the more likely it is that changes will happen. Examples are as follows:

* Alterations to drawings – this can happen as a result of several different factors. The architect or the client may decide at a fairly late stage that changes need to be made to the design of the project. This will require all documents that rely on information from the drawings to be amended. This could mean changes to the schedule, specification and work programmes and the need for materials and labour at particular times.

* Variation to contracts – although the construction company may have agreed with the client to carry out work based on particular drawings and specifications, changes to design and to the requirements may happen. It may be necessary to put in new estimates for additional work and to inform the client of any likely delays.

* Changes to risk assessments – it is not always possible to predict exactly what hazards will be encountered during a project. Neither is it possible to predict whether new legislation will come into force that requires extra risk assessments.

* Work restrictions – although the site will have been surveyed for access and cleared of obstacles such as low trees, problems may arise during the work. Local residents, for example, may complain about lighting and noise. This could reduce working hours on site. This could all have an impact on the schedule of work.

* Change in circumstances – this could cover a wide variety of different problems. Key suppliers may not be able to deliver materials or components on time. Tried and trusted sub-contractors may not be available. The client may run out of money or a problem may be unearthed during excavation and preparation of the site.

REED TIP

Open and frank communication means being able to say no if something is not possible. It's OK to say that something can't be done, rather than saying 'yes, yes, yes' and then being unable to complete a task.

DID YOU KNOW?

During the construction of the Olympic basketball site in London the whole site had to be evacuated when a Second World War bomb was found. It had to be removed by specialists before work could continue.

TEST YOURSELF

1. What is the system that is gradually taking over from CAD as the main way to produce construction drawings?

   a. CTIBM

   b. CIM

   c. BIM

   d. SIM

2. What does a block plan show?

   a. The construction site and its surrounding area

   b. Local boundaries and roads

   c. Elements and components

   d. Constructional details

3. Who might give you a delivery note?

   a. A postal worker

   b. An architect

   c. A contractor

   d. A supplier

4. Which type of construction drawings show the different faces or views of an object?

   a. Orthographic

   b. Section

   c. Elevation

   d. Plan

5. How many different types of pictorial projection are there?

   a. 3

   b. 4

   c. 5

   d. 6

6. Which of the following is usually a bid for a fixed amount of work in competition with other companies?

   a. Estimate

   b. Quotation

   c. Tender

   d. Invoice

7. To calculate the area of a room, which two measurements are needed?

   a. Length and height

   b. Height and width

   c. Length and width

   d. Length and circumference

8. If you had 5 workmen being paid £25 per hour and they were working for 4 hours, what would be the total labour cost?

   a. £125

   b. £250

   c. £500

   d. £600

9. Some contracts state that if a deadline is missed a fine has to be paid. What are these called?

   a. Terms and conditions

   b. Wastage

   c. Penalty clause

   d. Critical date payment

10. A business's operating profit is its gross profit minus which of the following?

    a. Tax

    b. Overheads

    c. Net profit

    d. Labour costs

Chapter 3

Unit CSA–L3Core08
ANALYSING THE CONSTRUCTION INDUSTRY AND BUILT ENVIRONMENT

LEARNING OUTCOMES

LO1: Understand the different activities undertaken within the construction industry and built environment

LO2: Understand the different roles and responsibilities undertaken within the construction industry and built environment

LO3: Understand the physical and environmental factors when undertaking a construction project

LO4: Understand how construction projects can benefit the built environment

LO5: Understand the principles of sustainability within the construction industry and built environment

INTRODUCTION

The aim of this unit is to:

* help you understand more about the construction industry and its place in society.

CONSTRUCTION INDUSTRY AND BUILT ENVIRONMENT ACTIVITIES

Half of all the non-renewable resources used across the globe are consumed by construction. Construction and the built environment are also linked with the pollution of drinking water, the production of waste and poor air quality.

Nevertherless, buildings create wealth. In the UK, buildings represent three-quarters of all wealth. Buildings are long-term assets. Today it is recognised that buildings should have the ability to satisfy user needs for extended periods of time. They must be able to cope with any changing environmental conditions. They also need to be capable of being adapted over time as designs and demands change.

There is an increasing move towards naturally lit and well-ventilated buildings. There is also a move towards buildings that use alternative energy sources.

The first part of this chapter looks at the type of work that has developed around the broader construction industry and built environment. It looks at the work that is undertaken and the different types of clients who use the construction industry.

Range of activities

The construction industry and the broader built environment is a highly complex network of different activities. While there are a great many small businesses that focus on one particular aspect of construction, they need to be seen as part of a far larger industry. Increasingly it is a global industry, with major business organisations operating not just in the UK but also in a wide variety of locations around the world. Their skills and expertise are in great demand wherever there is construction. The following table outlines some of the activities that are undertaken by the construction industry and the broader built environment.

Activity	Description
Building	This is the accepted and traditional activity of the construction industry. It involves building homes and other structures, from garden walls to entire housing estates or even Olympic villages.
Finishing	Finishing refers to a part of the industry that focuses on decorative work, such as painting and decorating. Once buildings are completed, in order to make them ready for habitation a broad range of professions are needed. Plumbers will install water and sanitation. Electricians will connect electrical services and equipment. Interior designers will create the desired look for the building.
Architecture	Architects and technicians design buildings for clients. The structures are designed to meet the needs of the client while ensuring that they conform to Building Regulations, local planning laws and decisions, as well as other legislation such as CDM Regulations and ensure they are sustainable.
Town planning	Town planning involves organising the broader built environment in a particular area. Town planners need to examine each planning application and see how it fits into the overall long-term future of the area. They need to ensure that the area meets the needs of future generations.
Surveying	This involves measuring and examining land on which building or other external work will take place. It can involve setting out the building. Surveyors use drawings by an architect to correctly position the building. They will be able to work out the area of the building and any volumes. Building surveyors check that the building is structurally sound, while quantity surveyors look after costs.
Civil engineering	Civil engineers are usually involved in major projects, such as road and railway building, the construction of dams, reservoirs and other projects that are not usually buildings. They are involved in what is known as infrastructure projects, such as transport links, networks and hubs.
Repair and maintenance	All buildings need professionals who are able to repair and maintain a broad range of features. From the foundations to the roof, carpenters, builders, electricians, plumbers and more specialist companies, such as pest control, can all be considered to be part of the repair and maintenance side of the industry. Pre-1919 buildings also have particular requirements and are maintained and repaired in a way that suits their construction and to avoid further damage or inappropriate work that will look out of place. This requires people who have specialist heritage skills.
Building engineering services	When buildings are occupied they need to be continually supported in terms of a rigorous checking and maintenance programme. This part of the industry can deal with lifts and escalators, lighting and heating, fire alarms and other inbuilt systems.
Facilities management	For larger commercial buildings or hospitals, schools, colleges and universities, systems need to be in place to replace parts of the building if they wear out or are damaged. This includes cleaning, air conditioning companies, painting and decorating, replacement of doors, windows and a host of other activities.
Construction site management	Construction sites can be complex and demanding places and someone needs to organise them and to monitor progress. Construction site management involves organising the delivery of materials, security, safety, the management of the workforce and contractors.
Plant maintenance and operation	Just as commercial buildings and dwellings need constant maintenance, so too do factories and other sites where products are made or processes are carried out. These individuals can be involved in the energy industry, at gas, oil and nuclear plants, or be responsible for maintaining factories that produce vehicles or food.
Demolition	Demolition experts are responsible for levelling sites in a safe and controlled way. They may have to demolish buildings that could contain asbestos or they may have to use controlled explosions.

Table 3.1

Types of work

There is also a wide range of work that is undertaken in different sectors. Some of this is very specialised work. Some companies will focus purely on that type of work, gaining a reputation and expertise in that area. The following table outlines the types of work that is undertaken within the construction industry.

Type of work	Description
Residential	This is any work connected with domestic housing or dwellings. It can include the building of new homes, extensions or renovations on existing homes and the construction of affordable accommodation for organisations such as housing associations.
Commercial	This is work related to any buildings used by businesses. It can include factories, office blocks, production units, industrial units or private hospitals.
Industrial	This is more specialist work, as it can involve construction, including civil engineering, of heavy industrial factories, such as oil refineries or plants for car manufacturing.
Retail	This can include building or refurbishing shops in high streets or the construction of out-of-town retail parks.
Recreational and leisure	Many of these projects are designed for use by communities, such as sports facilities, fitness clubs, leisure centres, swimming pools and other community sports projects. In the past decade the construction industry was involved in the various London 2012 Olympic facilities.
Health	This includes specialist building services to create hospitals and other health facilities, such as doctors' surgeries and care homes.
Transport infrastructure	This is another broad area of work that includes roads, motorways, bridges, railways, underground trains and tram systems, as well as airports, bus routes and cycle paths.
Public buildings	This is the building and maintenance of large buildings for local and central government. It can include offices, town halls, art galleries, museums and libraries.
Heritage	Heritage involves work on listed properties of historical importance. This is a specialist area, as Building Regulations, planning laws and Listed Status require any work to be carried out in sympathy with the original design of the building.
Conservation	This is an increasingly important area of work, as it involves the protection of natural habitats. It would involve work in National Parks, Areas of Outstanding Natural Beauty, animal sanctuaries and could also include construction work related to coastal erosion and flood defences.
Educational	This is the construction of schools, colleges, universities and other buildings used for educational purposes.
Utilities and services	This is work that is related to the installation, maintenance and repair of the key utilities, which include gas, electricity and water.

Table 3.2

Types of client

As we have seen, there is a huge range of activities and types of work in the broader construction industry and built environment areas. This means that there is a huge range of different potential types of client. Some are private individuals but at the other end of the scale they might be huge companies or government departments. The following table outlines the range of different types of client.

Type of client	Description
Private	These are usually individual owners of homes or buildings. They may be people who want work done on their own homes, or on their own business premises, such as a small shop. Many of the individual shop or business owners may be sole traders. These are individuals who run and own a small business.
Corporate	Corporate is a term that is used to describe larger companies or businesses. They can be individuals that run factories, larger shops, industrial units or some kind of service-based organisation, including banks, insurance companies and estate agents. Some construction companies have long-term contracts with corporate businesses, which have many branches around the country. There is a rolling programme of maintenance, upgrading and repair. The companies can be public limited companies (PLC), who are owned by shareholders with their shares traded on the Stock Exchange.
Government	The government can be a client on a local, regional or national basis. This part of the industry has become more complicated, as there are multiple levels of government across the UK. There is also a Scottish Parliament and a Welsh Assembly, in addition to the UK Parliament based in London. Local councils will be responsible for maintaining a wide range of services and they will also be involved in construction. This includes schools, roads, the maintenance of social housing and parks and leisure facilities. In addition to this there are government departments based in London with regional offices, such as the Ministry of Defence, which is responsible for facilities related to the armed forces, and the National Health Service, which is responsible for hospitals and other health provision. The government (including local authorities and non-departmental public bodies) must comply with strict procurement (buying) rules, which often involve tenders. They also have limited budgets, which could affect the building project's schedule.

Table 3.3

CONSTRUCTION INDUSTRY AND BUILT ENVIRONMENT ROLES AND RESPONSIBILITIES

As we have discussed, the construction industry and the built environment is a complex network of different activities. As the industry has developed over time it has become important for individuals to specialise and take on specific roles and responsibilities.

Roles and responsibilities of the construction workforce

The following tables show the broad range of different roles and briefly outline their responsibilities within the construction industry.

From the design and planning phase onwards

Role	Responsibilities
Client	The client, such as a local authority, commissions the job. They define the scope of the work and agree on the timescale and schedule of payments.
Customer	For domestic dwellings, the customer may be the same as the client, but for larger projects a customer may be the end user of the building, such as a tenant renting local authority housing or a business renting an office. These individuals are most affected by any work on site. They should be considered and informed with a view to them suffering as little disruption as possible.
Architect	They are involved in designing new buildings, extensions and alterations. They work closely with clients and customers to ensure the designs match their needs. They also work closely with other construction professionals, such as surveyors and engineers.
Estimator	Estimators calculate detailed cost breakdowns of work based on specifications provided by the architect and main contractor. They work out the quantity and costs of all building materials, plant required and labour costs.
Planner	Consultant planners such as civil engineers work with clients to plan, manage, design or supervise construction projects. There are many different types of consultant, all with particular specialisms.
Buyer	This individual works closely with the quantity surveyor. It is the buyer's job to source suitable materials as specified by the architect. They will negotiate prices and delivery dates with a range of suppliers.

Table 3.4

Surveying

Role	Responsibilities
Land agent	This is an individual who is authorised to act as an agent in the sale of land or buildings by the owner. Basically they are estate agents that sell plots of land.
Land surveyor	A land surveyor measures, records and then produces a drawing of the landscape. The data that they produce is used to plan out construction work.
Building surveyor	A building surveyor is responsible for making sure that both old and new buildings are structurally sound. They are involved in the design, maintenance, repair, alteration and refurbishment of buildings.
Quantity surveyor	Quantity surveyors are concerned with building costs. They balance maintaining standards and quality against minimising the costs of any project. They need to make choices in line with Building Regulations. They may work either for the client or for the contractor.

Table 3.5

Engineering

Role	Responsibilities
Building services engineers	They are involved in the design, installation and maintenance of heating, water, electrics, lighting, gas and communications. They work either for the main contractor or the architect and give instruction to building services operatives.
Structural engineer	Structural engineers are involved in ensuring that construction work is strong enough to deal with its use and the external environment. So they will be involved in the shape, design and the materials used. They will not only deal with new construction work but also advise on older buildings or buildings that have been damaged.
Consulting/building engineer	These individuals are involved in site investigation, building inspection and surveys. They get involved in a wide range of construction and maintenance projects.
Plant engineer	A plant engineer is responsible for maintaining and repairing a variety of machinery and equipment. They will also install and modify machinery and equipment in factories as part of an industrial or manufacturing process.
Site engineer	A site engineer is involved in setting out the plans for sewers, drains, roads and other services.
Specialist engineer	A good example of a specialist engineer is one that deals entirely with insulation. They will advise and install a range of energy conservation materials and equipment. A geotechnical engineer is another example. They carry out investigations into below foundation level and look at rock, soil and water.
Mechanical engineer	Mechanical engineers are primarily involved in installing and maintaining machinery and tools. It is a wide ranging profession but they will have overall responsibility for their particular area of work.
Demolition engineer	These engineers perform the task of tearing down old structures or levelling ground to make way for new buildings.
Infrastructure engineer	These engineers deal with the planning, construction and management of roads, bridges and similar structures.

Table 3.6

REED TIP

Any work experience is relevant to your job applications. It doesn't have to be paid work – e.g. volunteering to help run Scout and Guide activities shows your sense of responsibility. Think of the times when others have had to rely on you.

CASE STUDY

South Tyneside Homes

South Tyneside Council's
Housing Company

Your apprenticeship is just the start

Gary Kirsop, Head of Property Services, started at South Tyneside Homes as an apprentice 24 years ago.

'After becoming qualified, I had two options. I could have stayed working on the sites and become a site manager or technical assistant. I qualified as a building surveyor, doing my advanced craft at Sunderland College. After that, I went to Newcastle College to do my ONC and CHND, and eventually went on to finish a degree at Newcastle.

When I was a technical assistant I worked on education and public buildings, and spent a year in housing. As a technical assistant I was working on drawing (CAD), estimating small jobs to large jobs. Then an opportunity for Assistant Contracts Manager on capital works came up. Since then, I've also worked in disrepair and litigation, as well as two years with the empty homes department, and I've worked as a Construction Services Manager, responsible for the capital side, new homes, decent homes, and the gas team.

Four years ago, the Head of Property Services job came up and it's been a fantastic opportunity – my team has been one of the best in the country for performance. My department is responsible for repairs and maintenance, capital works, empty homes, and management of the operational side. We do responsive repairs for emergency situations, planned repairs, work for the "Decent Homes" programme where we bring properties up to standard, and we've recently built four new bungalows. Anything in construction, we have the skills and labour to do it in property services.

The full management team here in property services all started as apprentices, like me. It really helps that we understand the whole process from beginning to end.

So you can see that doing your apprenticeship is not only great in itself, but it also gives you skills for life and ongoing opportunities for education, training and your career.'

PHYSICAL AND ENVIRONMENTAL FACTORS AND CONSTRUCTION PROJECTS

Increasingly, people working in construction and the built environment are being asked to ensure that they minimise physical and environmental impacts when carrying out construction work. Construction has an enormous impact on the environment. Environmental measures will depend on the nature of the work and the site. For example, excavations that result in changes in the levels of land can cause problems with water quality and soil erosion. Many of these negative impacts can be reduced during the planning stage.

Physical and environmental factors

Physical factors relate to the impact that any new construction project will have on any existing structures and their occupants. Any new construction project is going to have a negative impact on home owners and businesses. There will be increased traffic on roads and a host of other considerations.

Once the construction has been completed there may be longer term impacts. A prime example would be building a new housing development in an area that lacks good roads, sufficient schools or access to health facilities. During the planning and development stage these factors will be looked at to see what the knock-on effects might be in the short and long term.

Environmental factors concern the impact that a construction project has on the natural environment. This would include any possible impacts on trees and vegetation, wildlife and habitats. It can also have an impact on the air quality or noise levels in the area.

Physical factors and the planning process

There is a wide variety of different physical factors that have to be taken into consideration during the planning process. These are outlined in the following table.

Physical factor	Explanation
Planning requirements	The majority of new developments or changes to existing buildings do require consent or planning permission. The local planning authority will make a decision whether any such construction will go ahead. Each authority has a development framework that outlines how planning is managed. This includes the change of use of a building or a piece of land.
Building Regulations	There are 14 technical parts of the Building Regulations covering everything from structural safety to electrical safety. They also outline standards of quality of work and materials used. All new developments and major changes to existing buildings must comply with Building Regulations.
Development or land restrictions	This is a complicated area, as there are often many restrictions on building and the use of land. One of the most complex is restrictive covenants, which are created in order to protect the interests of neighbours. They might restrict the use of the land and the amount of building work that can take place.
Building design and footprint	The footprint is the physical amount of space or area that the proposed development takes up on a given plot of land. There may be limits as to the size of this footprint. In terms of building design, certain areas may have restrictions as the local authority may not approve the construction of a building that is out of character, or that would adversely affect the overall look of the area.
Use of building or structure	Each building or structure will have a Use Class, such as 'residential', 'shops' or 'businesses'. Redeveloping an existing building and not changing the use to which it is put, for example renovating a building from a butcher to a chemist, does not usually require planning permission. However, changing from a bank to a bar would require planning permission. Certain uses, due to their unique nature, do not fall into any particular Use Class and planning permission is always required. A good example would be a nightclub or a casino.

Physical factor	Explanation
Impact on local amenities	During the construction phase it is likely that roads or access may have to be blocked, which could impact on local businesses. In the longer term additional traffic and the need for parking may have an impact on local amenities, as will the demand for their use.
Impact on existing services and utilities	Any new development or major change in use of an existing structure may put extra strain on services and utilities in the area. A new housing development, for example, would require power cables to be run to the site. It would also need excavation work to connect it to the sewers and underground pipes run onto the site for potable water. All of this is potentially disruptive and may require considerable investment by the utility or service provider.
Impact on transportation infrastructure	Major new developments will have a huge impact on the roads and public transport in an area. Permission for major developments often comes with the requirement to improve access routes, build new roads and the requirement to make a contribution to improvements in the infrastructure. New developments can radically change the flow of traffic in an area and may have a knock-on effect in terms of maintenance and repair in the longer term.
Topography of the proposed development site	The term topography refers to the location of the site and how dominant it will be in the local landscape. Obviously a development that is situated on a hill or ridge is far more obvious and will have a longer lasting impact on the local area. If the development is considered to be too obtrusive or visible then it may be deemed as inappropriate to situate the development on that site.
Greenfield or brownfield site	A greenfield site is an area of land that has never been used for non-agricultural purposes. A brownfield site is usually former industrial land, or land that has been used for some other purpose and is no longer in use. There is more information on greenfield and brownfield sites in the next section of this chapter.

Table 3.7

Environmental factors and the planning process

Just as there are physical factors, there are also different environmental factors that need to be considered. Some of the major ones are detailed in the following table.

Environmental factor	Explanation
Topography of the development site	As mentioned in the previous table, the topography of the development site can have a marked impact on the local environment. It may dominate what is otherwise a predominantly natural environment, perhaps with woodland or rolling hills.
Existing trees and vegetation	Sites may have to be cleared in order to provide the necessary space for the footprint of the structure. It may be prohibited to remove or otherwise interfere with certain trees and vegetation, as they may be protected. The normal course of events is to minimise the impact on existing plant life and to have a replanting phase after the site has been developed.
Impact on existing wildlife and habitats	Any potential impact on wildlife and plants that are under threat could mean that the site would not receive the go ahead. An environmental impact study will identify whether there are any specific dangers that will affect the natural habitat of the area, or endanger any local species of wildlife.
Size of land and building footprint	There is a formula that determines the usually permitted footprint of a piece of land compared to the actual size of the plot of land. For example, a 4-bedroom house on an average housing estate would take up approximately 1/12th of an acre (11.5 m × 29 m).

Environmental factor	Explanation
Access to the building or structure	It is not only the building plot that needs to be considered in terms of its environmental impact. Access to the site is another concern. Existing roads may have to be widened, perhaps a roundabout installed. Alternatively new roads may have to be built across other plots of land. For pedestrian traffic footpaths may also be necessary. These can either be alongside existing roads or built alongside new roads, requiring even more space. There may be existing footpaths and this could mean that access needs to be provided through the site or the footpaths diverted.
Supply of services to the building or structure	Running above ground services and utilities to the site may also present a problem as far as its impact on the environment is concerned. It may not be possible to allow features such as pylons or street lights to dominate the landscape.
Natural water resources	New developments can affect the biodiversity of an area by impacting on natural waterways. Local wildlife and plants rely on this resource. In addition to this, construction could either pollute or affect the quality of the local water.
Land restrictions	There may be land restrictions that limit either the use or the size of any development. Developments will not be allowed to adversely affect surrounding properties and owners. There are conservation areas, scheduled monuments, archaeological sites and scheduled or listed buildings. These are all protected and construction on or near them is either prohibited or severely limited.
Future development and expansion	Although the intention may be to restrict the environmental impact of the site in the first phase of development, in the future this might not be possible. Major housing development is often carried out in phases and the size of the development will gradually increase as demand increases. It is therefore important when permission is initially given that the likelihood of future development and expansion is taken into account.

Table 3.8

Figure 3.1 Trees on a proposed site may need to be protected during construction work

DID YOU KNOW?

In some cases, Tree Preservation Orders are put in place by the local planning authority. These prevent the removal of trees or work on them without permission. Some land, due to its natural beauty, importance to local wildlife and plants or special geological features, can also be protected, making it impossible for any development to take place on the site.

HOW CONSTRUCTION PROJECTS BENEFIT THE BUILT ENVIRONMENT

The construction industry is one of the UK's largest employers. It is a hugely diverse industry. Construction projects can have a massive impact on the built environment. They can rejuvenate whole areas; improve the housing stock, amenities and the general life and well-being of the local population. The built environment describes the overall look and layout of a specific area. Each new construction project and its architectural design will have an impact on that built environment and the broader, natural environment. If it is carefully and sympathetically planned and organised it can have a positive impact on the way people live, work and interact with one another.

Each new development has enormous environmental, social and economic consequences. Increasingly it has a role to play in ensuring that our built environment has a strong and sustainable future.

Land types available for development and their advantages and disadvantages

In March 2012 the National Planning Policy Framework was published, which aims to review planning guidance across the UK. The idea was to encourage the building of domestic dwellings. It stated that there would be a policy to try to use as many brownfield sites as possible, but that greenfield sites in rural areas would no longer be protected at any cost. Where development was necessary it would take place, as there was a huge demand for homes, shops and workplaces.

The first targets for development would be sites that had been used in the past for other purposes.

Greenfield land or sites

Greenfield sites are usually either agricultural or amenity land. Given the fact that there is a housing crisis in the UK and that land needs to be allocated to build millions of new homes, greenfield sites are very much under consideration.

The problem in doing this is that there is huge resistance, particularly in rural areas, to losing greenfield sites for the following reasons:

* Once a greenfield site has been developed it is extremely unlikely that it will ever return to agricultural use. Any loss of agricultural land means a reduction in the amount of food that can be produced in the UK. There might also be a drop in employment in the local area as fewer farm workers are needed.

* Natural habitats of wildlife and plants are destroyed forever.

* Greenfield or amenity land, if lost, means that the land can no longer be used for leisure and recreation.

* Developments on greenfield sites can have a negative impact on the local transport infrastructure and will increase the amount of energy used because things are further away from town centres.

* The loss of green belts of agricultural land around cities, towns and villages means that each separate area loses its identity and in effect becomes a suburb of a larger town or city.

Figure 3.2 Building on greenfield and greenbelt land is a controversial issue

Brownfield land or sites

Brownfield sites are pieces of land that have been previously developed. They were probably used for either industrial or commercial purposes, but are now derelict and abandoned.

Figure 3.3 Brownfield sites have already been built on

Brownfield sites can be found in areas where there is a high demand for new homes. It has been estimated that there are more than 66,000 hectares of brownfield sites in England alone. At least a third of this land can be found in the southeast of England, where there is the highest demand for housing. Around 60 per cent of new housing is being built on brownfield sites. This is a trend that is likely to accelerate over the next 10 years.

Brownfield sites are not just used for housing projects but are also sites for commercial buildings, as well as recreational sites and newly planted woodland.

Reclaimed land

There are areas, particularly around the coast and in estuaries, which for many years have been bogs or salt marshes. These damp grasslands can be gradually drained of water and eventually provide agricultural land or, in some cases, land suitable for housing developments. With global warming and climate change threatening to permanently flood huge areas of the UK, it may seem strange to consider humans reversing the process.

The area is converted by digging flood relief channels and drainage ditches to encourage the water to flow out and away from the land. To protect the land during this process banks are built to keep out river and seawater. It is a long and involved process but can provide possible land for redevelopment. This process has been successful in many different parts of the world, notably in the Fens in East Anglia, on the Netherlands coast, where pumping stations reclaim land from the sea, and in the Middle and Far East where huge projects have reclaimed vast areas of land.

Figure 3.4 Reclaiming land enables it to be put other uses

Contaminated land

Many brownfield sites, particularly those once used for industrial purposes, are contaminated with varying levels of hazardous waste and pollutants. Before any development can take place an environmental consultant will organise the analysis of soil, ground water and surface water to identify any risks.

Special licences are required to reclaim brownfield sites and this can be a very expensive process for developers. The main way of dealing with brownfield sites is a process known as remediation. This involves the removal of any known contaminants to a level that will not affect the health of anyone living or working on the site both during construction and after building is complete.

Not all brownfield sites are, therefore, suitable or cost-effective. In some cases the cost of removing the contaminants exceeds the value of the land after it has been developed. There are new ways of dealing with contaminants:

* Bioremediation – this uses bacteria, plants, fungi and micro-organisms to destroy or neutralise contaminants.

* Phytoremediation – plants are encouraged to grow on the site and the contaminants are taken up into the plant and stored in their leaves and stems.

* Chemical oxidation – this involves injecting oxygen or oxidants into contaminated soil and water to destroy contaminants.

DID YOU KNOW?

Brownfield redevelopment has huge advantages as it not only deals with environmental health hazards, but also regenerates areas. It can provide affordable housing, jobs and conservation.

Figure 3.5 Contaminated land must be cleaned before use

Social benefits of construction development

The construction industry and the built environment do provide a range of potential benefits, particularly to local areas. These are examined in the following table.

Social benefit	Explanation
Regeneration of brownfield sites	Disused land, usually former industrial sites, and have been developed for new housing and commercial sites. In London, virtually the whole of the 2012 Olympic village was built on brownfield sites.
Local employment	Construction sites need the skills of local construction workers and offer opportunities for small businesses. Long-term projects offer long-term employment for local people.
Improved housing	New developments and refurbishment of older properties provide greener and more energy efficient dwellings. This has a long-term positive impact for the environment and the reduction in the use of non-renewable resources.
Improvements to local infrastructure	A new development of any size often comes with the requirement for the developers to contribute towards the building of new roads and other infrastructure projects for the area. New developments, in order to work, need access roads, transport and other facilities.
Improvements to local amenities	Modern housing developments and commercial properties need to have amenities near them in order to make them viable in the longer term. This means the building of schools, hospitals, health centres and shops.

Table 3.9

DID YOU KNOW?

In December 2012 work to transform 53 acres (21ha) of urban wasteland into an innovative £100m housing development of up 800 new homes got under way in Cardiff. It is situated at the site of an old paper mill on the banks of the River Ely. The development provides much-needed affordable housing for the city.

Figure 3.6 Sustainable developments aim to be pleasant places to live

SUSTAINABILITY

Carbon is present in all fossil fuels, such as coal or natural gas. Burning fossil fuels releases carbon dioxide, which is a greenhouse gas linked to climate change.

Energy conservation aims to reduce the amount of carbon dioxide in the atmosphere. The idea is to do this by making buildings better insulated and, at the same time, making heating appliances more efficient. It also means attempting to generate energy using renewable and/or low or zero carbon methods.

According to the government's Environment Agency, sustainable construction is all about using resources in the most efficient way. It also means cutting down on waste on site and reducing the amount of materials that have to be disposed of and put into **landfill.**

In order to achieve sustainable construction the Environment Agency recommends:

- reducing construction, demolition and excavation waste that needs to go to landfill
- cutting back on carbon emissions from construction transport and machinery
- responsibly sourcing materials
- cutting back on the amount of water that is wasted
- making sure construction does not have an impact on **biodiversity.**

KEY TERMS

Landfill

– 170 million tonnes of waste from homes and businesses are generated in England and Wales each year. Much of this has to be taken to a site to be buried.

Biodiversity

– wherever there is construction there is a danger that the wildlife and plants could be disturbed or destroyed. Protecting biodiversity ensures that at risk species are conserved.

What is meant by sustainability?

In the past buildings have been constructed as quickly as possible and at the lowest cost. More recently the idea of sustainable construction has focused on ensuring that the building is not only of good quality and that it is affordable, but that it is also energy efficient.

Sustainable construction also means having the least negative environmental impact. So this means minimising the use of raw materials, energy, land and water. This is not only during the build period but also for the lifetime of the building.

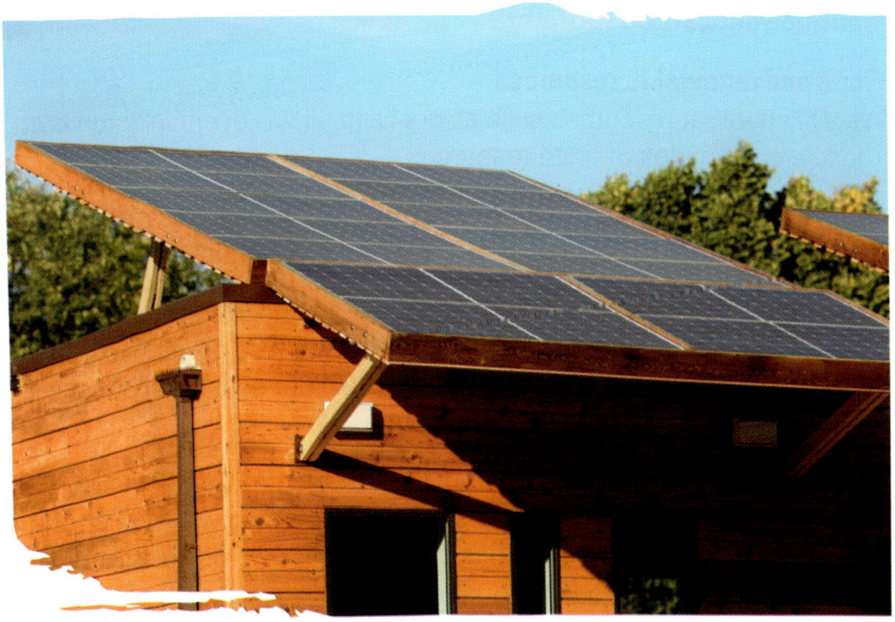

Figure 3.7 Eco houses are becoming more common

Construction and the environment

In 2010, construction, demolition and excavation produced 20 million tonnes of waste that had to go into landfill. The construction industry is also responsible for most illegal fly tipping (illegally dumping waste). In any year there are at least 350 serious pollution incidents caused as a result of construction.

Figure 3.8 Always dispose of waste responsibly

Regardless of the size of the construction job, everyone in construction is responsible for the impact they have on the environment. Good site layout, planning and management can help reduce this impact.

Sustainable construction helps to encourage this because it means managing resources in a more efficient way, reducing waste and reducing your **carbon footprint.**

Finite and renewable resources

We all know that resources such as coal and oil will eventually run out. These are examples of finite resources.

Oil is not just used as fuel – it is used in plastic, dyes, lubricants and textiles. All of these are used in the construction process.

Renewable resources are those that can be produced by moving water, the sun or the wind. Materials that come from plants, such as biodiesel, or the oils used to make some pressure-sensitive adhesives, are examples of renewable resources.

The construction process itself is only part of the problem. It is also the longer term impact and demands that the building will have on the environment. This is why there has been a drive towards sustainable homes and there is a Code for Sustainable Homes, which is a certification of sustainability for new build housing.

The future

Sustainability also means ensuring that future generations do not suffer from the ill-considered activities of today's generation. The following table outlines some of the present dangers and concerns.

Present or future concern	Explanation
Global shortages	Many naturally found resources will eventually run out and they will have to be replaced with alternatives. Acting now to discover, develop or use alternatives will delay this. Construction is at the forefront of finding alternatives and looking at different construction materials and methods.
Needs of future generations	Buildings constructed today must to be useful and affordable for future generations. At the same time, materials and construction methods should not leave a bad legacy that future generations have to deal with.
Global warming	The construction industry has been criticised over its contribution to global warming. A lack of co-ordination between different parts of the industry has produced poor quality, energy-inefficient buildings. The government is keen to ensure that the industry trains people about the principles of sustainable design and efficient technologies. These steps need to be put in place to inform decisions at the design stage of a building.
Climate change	Construction projects need to take into account the effects of climate change and consider ways to reduce the project's impact on the environment. This means minimising carbon emissions, using sustainable (or renewable) energy and reducing water consumption.
Extinction of species and vegetation	Global warming and climate change has an impact on animals and plants. On a local level, this is also a problem as construction can destroy natural habitats. Increasingly, this is closely monitored and environmental impact studies are used to prevent this from happening.
Destruction of natural resources	There are strict planning laws that aim to prevent the industry from destroying or harming natural habitats. Ancient woodland, sites of scientific importance and other sites of interest are all protected. It is also the case that development in areas that are likely to flood or cause flooding are prohibited or controlled.

Table 3.10

KEY TERMS

Global warming

– a rise in temperature of the earth's atmosphere. The planet is naturally warmed by rays, some being reflected back out into space. The atmosphere is made up of gases (some are called greenhouse gases) which are mainly natural and form a kind of thermal blanket. The human-made gases are believed to make this blanket thicker, so less of the heat escapes back into space. Over the past 100 years, our climate has seen some rapid changes. This is believed to be linked to changes in the makeup of the atmosphere and land use.

Climate change

– the burning of fossil fuels (coal, gas, wood, oil) has resulted in an increase in the amount of greenhouse gases. This has pushed up global temperatures. Across the world, millions do not have enough water, species are dying out and sea levels are rising. In the UK we see extreme events such as flooding, storms, sea level rise and droughts. We have wetter warmer winters and hotter drier summers.

Figure 3.9 Climate change may be a serious problem over the next decades

Social regeneration

Construction projects are often used to regenerate areas of the UK that have lacked investment in the past. As industry develops and changes over time, whole areas that would once have been extremely busy in the past now have empty industrial units and high unemployment levels. As the area loses jobs housing deteriorates, as does the local infrastructure, as there is no money in the local economy.

Redeveloping these waste sites is seen as a way in which a whole area can be regenerated or reborn. Construction projects bring jobs relating to the project but they also bring the promise of longer term jobs. These areas have relatively cheap land and lower rents. Also the workforce expects lower rates of pay. This attracts businesses to relocate to the new buildings created by construction developers. This brings work, improved housing, and improvements to the local infrastructure and amenities.

Sustainability and its benefits

Energy efficiency is all about using less energy to provide the same result. The plan is to try to cut the world's energy needs by 30 per cent before 2050. This means producing more energy efficient buildings. It also means using energy efficient methods to produce the materials and resources needed to construct buildings.

Alternative methods of building

The most common type of construction in the UK is brick and blockwork. However there are plenty of other options:

* timber frame – using pre-fabricated timber frames which are then clad

* insulated concrete formwork – where a polystyrene mould is filled with reinforced concrete

* structural insulated panels – where buildings are made up of rigid building boards rather like huge sandwiches

* modular construction – this uses similar materials and techniques to standard construction, but the units are built off site and transported ready-constructed to their location.

Figure 3.10 Insulated concrete formwork

Figure 3.11 Modular construction

There are alternatives to traditional flooring and roofing, all of which are greener and more sustainable. Green roofing (both living roofs and roofs made from recycled materials) has become an increasing trend in recent years. Metal roofs made of steel, aluminium and copper often use a high percentage of recycled material. They are also lightweight. Solar roof shingles, or solar roof laminates, while expensive, decrease the cost of electricity and heating for the dwelling. Some buildings even have a living roof which consists of a waterproof membrane, a drainage layer, a growing material and plants such as sedum. This provides additional insulation, absorbs air pollution, helps to collect and process rainwater and keeps the roof surface temperature down.

Just as roofs are becoming greener, so too are the options for flooring. The use of renewable resources such as bamboo, eucalyptus and cork is becoming more common. A new version of linoleum has been developed with **biodegradable**, **organic** ingredients. Some buildings are also using floorboards and joists made from non-timber materials that can be coloured, stained or patterned.

Figure 3.12 Solar roof tiles provide their own solar power

Figure 3.13 A stained concrete floor can be a striking feature

KEY TERMS

Biodegradable

– the material will more easily break down when it is no longer needed. This breaking down process is done by micro-organisms.

Organic

– natural substance, usually extracted from plants.

An increasing trend has been for what is known as off-site manufacture (OSM). European businesses, particularly those in Germany, have built over 100,000 houses. The entire house is manufactured in a factory and then assembled on site. Walls, floors, roofs, windows and doors with built-in electrics and plumbing all arrive on a lorry. Some manufacturers even offer completely finished dwellings, including carpets and curtains. Many of these modular buildings are designed to be far more energy efficient than traditional brick and block constructions. Many come ready fitted with heat pumps, solar panels and triple-glazed windows.

Figure 3.14 A timber-framed HUF haus is assembled off site

Architecture and design

The Code for Sustainable Homes Rating Scheme was introduced in 2007. Many local authorities have instructed their planning departments to encourage sustainable development. This begins with the work of the architect who designs the building.

Local authorities ask that architects and building designers:

* ensure the land is safe for development – that if it is contaminated this is dealt with first

* ensure access to and protection for the natural environment – this supports biodiversity and tries to create open spaces for local people

* reduce the negative impact on the local environment – buildings should keep noise, air, light and water pollution down to a minimum

* conserve natural resources and cut back carbon emissions – this covers energy, materials and water

* ensure comfort and security – good access, close to public transport, safe parking and protection against flooding.

Figure 3.15 Eco developments, like this one in London, are becoming more common

Using locally managed resources

The construction industry imports nearly 6 million cubic metres of sawn wood each year. Around 80 per cent of all the softwood used in construction comes from Scandinavia or Russia. Another 15 per cent comes from the rest of Europe, or even North America. The remaining 5 per cent comes from tropical countries, and is usually sourced from sustainable forests. However there is plenty of scope to use the many millions of cubic metres of timber produced in managed forests, particularly in Scotland.

Local timber can be used for a wide variety of different construction projects:

* Softwood – including pines, firs, larch and spruce – for panels, decking, fencing and internal flooring.

* Hardwood – including oak, chestnut, ash, beech and sycamore – for a wide variety of internal joinery.

Eco-friendly, sustainable manufactured products and environmentally resourced timber

There are now many suppliers that offer sustainable building materials as a green alternative. Tiles, for example, can be made from recycled plastic bottles and stone particles.

There is a National Green Specification database of all environmentally friendly building materials. This provides a checklist where it is possible to compare specifications of environmentally friendly materials to those of traditionally manufactured products, such as bricks.

Simple changes to construction, such as using timber or ethylene-based plastics instead of PVCU window frames is a good example.

Finding locally managed resources such as timber makes sense in terms of cost and in terms of protecting the environment.

PRACTICAL TIP

www.recycledproducts. org,uk has a long list of recycled surfacing products, such as tiles, recycled wood and paving and detials of local suppliers

DID YOU KNOW?

The Forest Stewardship Council has a system that verifies that wood comes from well-managed forests. The Programme for the Endorsement of Forest Certification Schemes promotes sustainable managed forests throughout the world. (www.fsc.org and www.pefc.co.uk)

The Timber Trade Federation produces a Timber Certification System. This ensures that wood products are labelled to show that they are produced in sustainable forests.

Building Regulations

In terms of energy conservation, the most important UK law is the Building Regulations 2010, particularly Part L. The Building Regulations:

- list the minimum efficiency requirements
- provide guidance on compliance, the main testing methods, installation and control
- cover both new dwellings and existing dwellings.

A key part of the regulations is the Standard Assessment Procedure (SAP), which measures or estimates the energy efficiency performance of buildings.

Local planning authorities also now require that all new developments generate at least 10 per cent of their energy from renewable sources. This means that each new project has to be assessed one at a time.

Energy conservation

By law, each local authority is required to reduce carbon dioxide emissions and to encourage the conservation of energy. This means that everyone has a responsibility in some way to conserve energy:

- Clients, along with building designers, are required to include energy efficient technology in the build.
- Contractors and sub-contractors have to follow these design guidelines. They also need to play a role in conserving energy and resources when actually working on site.
- Suppliers of products are required by law to provide information on energy consumption.

DID YOU KNOW?

The Energy Saving Trust has lots of information about construction. Its website is www.energysavingtrust.org.uk.

In addition, new energy efficiency schemes and building regulations cover the energy performance of buildings. Each new build is required to have an Energy Performance Certificate. This rates a building's energy efficiency from A (which is very efficient) to G (which is least efficient).

Some building designers have also begun to adopt other voluntary ways of attempting to protect the environment. These include: BREEAM (Building Research Establishment Environmental Assessment Method, a voluntary measurement rating for green buildings) and the Code for Sustainable Homes (a certification of sustainability for new builds).

energy® saving trust

Figure 3.16 The Energy Saving Trust encourages builders to use less wasteful building techniques and more energy efficient construction

High, low and zero carbon

When we look at energy sources, we consider their environmental impact in terms of how much carbon dioxide they release. Accordingly, energy sources can be split into three different groups:

* high carbon – those that release a lot of carbon dioxide

* low carbon – those that release some carbon dioxide

* zero carbon – those that do not release any carbon dioxide.

Some examples of high carbon, low carbon and zero carbon energy sources are given in the tables below.

High carbon energy source	Description
Natural gas or LPG	Piped natural gas or liquid petroleum gas stored in bottles
Fuel oils	Domestic fuel oil, such as diesel
Solid fuels	Coal, coke and peat
Electricity	Generated from non-renewable sources, such as coal-fired power stations

Table 3.11

Low carbon energy source	Description
Solar thermal	Panels used to capture energy from the sun to heat water
Solid fuel	Biomass such as logs, wood chips and pellets
Hydrogen fuel cells	Converts chemical energy into electrical energy
Heat pumps	Devices that convert low temperature heat into higher temperature heat
Combined heat and power (CHP)	Generates electricity as well as heat for water and space heating
Combined cooling, heat and power (CCHP)	A variation on CHP that also provides a basic air conditioning system

Table 3.12

Zero carbon energy	Description
Electricity/wind	Uses natural wind resources to generate electrical energy
Electricity/tidal	Uses wave power to generate electrical energy
Hydroelectric	Uses the natural flow of rivers and streams to generate electrical energy
Solar photovoltaic	Uses solar cells to convert light energy from the sun into electricity

Table 3.13

It is important to try to conserve non-renewable energy so that there will be sufficient fuel for the future. The idea is that the fuel should last as long as is necessary to completely replace it with renewable sources, such as wind or solar energy.

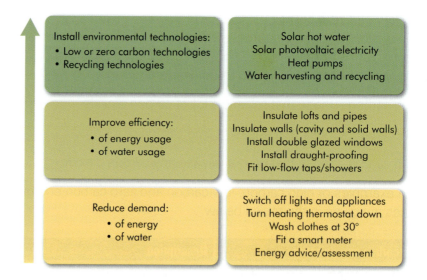

Install environmental technologies: • Low or zero carbon technologies • Recycling technologies	Solar hot water Solar photovoltaic electricity Heat pumps Water harvesting and recycling
Improve efficiency: • of energy usage • of water usage	Insulate lofts and pipes Insulate walls (cavity and solid walls) Install double glazed windows Install draught-proofing Fit low-flow taps/showers
Reduce demand: • of energy • of water	Switch off lights and appliances Turn heating thermostat down Wash clothes at 30° Fit a smart meter Energy advice/assessment

Figure 3.17 Working towards reducing carbon emissions

Alternative energy sources

There are several new ways in which we can harness the power of water, the sun and the wind to provide us with new heating sources. All of these systems are considered to be far more energy efficient than traditional heating systems, which rely on gas, oil, electricity or other fossil fuels.

Solar thermal

At the heart of this system is the solar collector, which is often referred to as a solar panel. The idea is that the collector absorbs the sun's energy, which is then converted into heat. This heat is then applied to the system's heat transfer fluid.

The system uses a differential temperature controller (DTC) that controls the system's circulating pump when solar energy is available and there is a demand for water to be heated.

In the UK, due to the lack of guaranteed solar energy, solar thermal hot water systems often have an auxiliary heat source, such as an immersion heater.

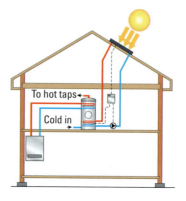

Figure 3.18 Solar thermal hot water system

Biomass (solid fuel)

Biomass stoves burn either pellets or logs. Some have integrated hoppers that transfer pellets to the burner. Biomass boilers are available for pellets, woodchips or logs. Most of them have automated systems to clean the heat exchanger surfaces. They can provide heat for domestic hot water and space heating.

Stove providing
room heat only

Stove providing
room heat and
domestic hot water

Stove providing
room heat, domestic
hot water and heating

Figure 3.19 Biomass stoves output options

Heat pumps

Heat pumps convert low temperature heat from air, ground or water sources to higher temperature heat. They can be used in ducted air or piped water **heat sink** systems.

There are different arrangements for each of the three main systems:

* Air source pumps operate at temperatures down to minus 20°C. They have units that receive incoming air through an inlet duct.

* Ground source pumps operate on **geothermal** ground heat. They use a sealed circuit collector loop, which is buried either vertically or horizontally underground.

* Water source pump systems can be used where there is a suitable water source, such as a pond or lake. Energy extracted from the water is used as heat.

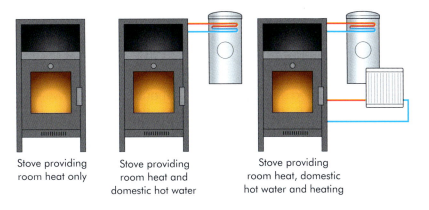

Figure 3.20 Heat pump input and output options

KEY TERMS

Heat sink

– this is a heat exchanger that transfers heat from one source into a fluid, such as in refrigeration, air conditioning or the radiator in a car.

Geothermal

– relating to the internal heat energy of the earth.

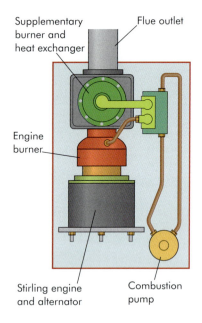

Supplementary burner and heat exchanger

Flue outlet

Engine burner

Stirling engine and alternator

Combustion pump

Figure 3.21 Example of a MCHP (micro combined heat and power) unit

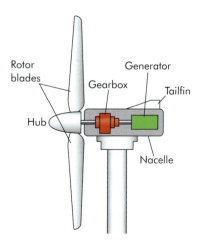

Rotor blades

Generator

Gearbox

Tailfin

Hub

Nacelle

Figure 3.22 A basic horizontal axis wind turbine

The heat pump system's efficiency relies on the temperature difference between the heat source and the heat sink. Special tank hot water cylinders are part of the system, giving a large surface-to-surface contact between the heating circuit water and the stored domestic hot water.

Combined heat and power (CHP) and combined cooling heat and power (CCHP) units

These are similar to heating system boilers, but they generate electricity as well as heat for hot water or space heating (or cooling). The heart of the system is an engine or gas turbine. The gas burner provides heat to the engine when there is a demand for heat. Electricity is generated along with sufficient energy to heat water and to provide space heating.

CCHP systems also incorporate the facility to cool spaces when necessary.

Wind turbines

Freestanding or building-mounted wind turbines capture the energy from wind to generate electrical energy. The wind passes across rotor blades of a turbine, which causes the hub to turn. The hub is connected by a shaft to a gearbox. This increases the speed of rotation. A high speed shaft is then connected to a generator that produces the electricity.

Solar photovoltaic systems

A solar photovoltaic system uses solar cells to convert light energy from the sun into electricity.

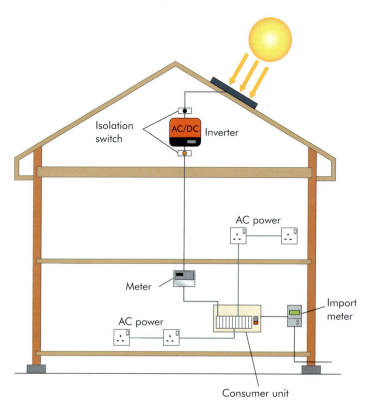

Isolation switch

AC/DC Inverter

AC power

Meter

AC power

Import meter

Consumer unit

Figure 3.23 A basic solar photovoltaic system

Energy ratings

Energy rating tables are used to measure the overall efficiency of a dwelling, with rating A being the most energy efficient and rating G the least energy efficient.

Alongside this, an environmental impact rating measures the dwelling's impact in terms of how much carbon dioxide it produces. Again, rating A is the highest, showing it has the least impact on the environment, and rating G is the lowest.

A Standard Assessment Procedure (SAP) is used to place the dwelling on the energy rating table. This will take into account:

* the date of construction, the type of construction and the location

* the heating system

* insulation (including cavity wall)

* double glazing.

The ratings are used by local authorities and other groups to assess the energy efficiency of new and old housing and must be provided when houses are sold.

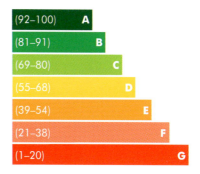

Figure 3.24 SAP energy efficiency rating table. The ranges in brackets show the percentage energy efficiency for each banding

Preventing heat loss

Most old buildings are under-insulated and will benefit from additional insulation, which can be for ceilings, walls or floors.

The measurement of heat loss in a building is known as the U Value. It measures how well parts of the building transfer heat. Low U Values represent high levels of insulation. U Values are becoming more important as they form the basis of energy and carbon reduction standards.

By 2016 all new housing is expected to be Net Zero Carbon. This means that the building should not be contributing to climate change.

Many of the guidelines are now part of Building Regulations (Part L). They cover:

* insulation requirements

* openings, such as doors and windows

* solar heating and other heating

* ventilation and air conditioning

* space heating controls

* lighting efficiency

* air tightness.

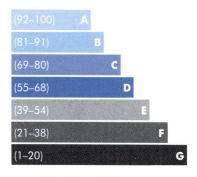

Figure 3.25 SAP environmental impact rating table

Building design

UK homes spend £2.4bn every year just on lighting. One of the ways of tackling this cost is to use energy saving lights, but also to maximise natural lighting. For the construction industry this means:

* increased window size

* orientating window angles to make the most of sunlight – south facing windows maximise sunlight in winter and limit overheating in the summer

* window design – with a variety of different types of opening to allow ventilation.

Solar tubes are another way of increasing light. These are small domes on the roof, which collect sunlight and direct it through a tube (which is reflective). It is then directed through a diffuser in the ceiling to spread light into the room.

Waste water recycling

Water is a precious resource, so it is vital not to waste it. To meet the current demand for water in the UK, it is essential to reduce the amount of water used and to recycle water where possible.

The construction industry can contribute to water conservation by effective plumbing design and through the installation of water efficient appliances and fittings. These include low or dual flush WCs, and taps and fittings with flow regulators and restrictors. In addition, rainwater harvesting and waste water recycling should be incorporated into design and construction wherever possible.

Statutory legislation for water wastage and misuse

Water efficiency and conservation laws aim to help deal with the increasing demand for water. Just how this is approached will depend on the type of property:

* For new builds, the Code for Sustainable Homes and Part G of the Building Regulations set new water efficiency targets.

* For existing buildings, Part G of the Building Regulations applies to all refurbishment projects where there is a major change of use.

* For owners of non-domestic buildings, tax reduction schemes and grants are available for water efficiency projects.

In addition, the Water Supply (Water Fittings) Regulations 1999 set a series of efficiency improvements for fittings used in toilets, showers and washing machines, etc.

Reducing water wastage

There are many different ways in which water wastage can be reduced, as shown in the table below.

Method	Explanation
Flow reducing valves	Water pressure is often higher than necessary. By reducing the pressure, less water is wasted when taps are left running.
Spray taps	Fixing one of these inserts can reduce water consumption by as much as 70 per cent.
Low volume flush WC	These reduce water use from 13 litres per flush to 6 litres for a full flush and 4 litres for a reduced flush.
Maintenance of terminal fittings and float valves	Dripping taps or badly adjusted float valves can cause enormous water wastage. A dripping tap can waste 5,000 litres a year.
Promoting user awareness	Users who are on a meter will certainly see a difference if water efficiency is improved, and their energy bills will be reduced if they use less hot water.

Table 3.14

Captured and recycled water systems

There are two variations of captured and recycled water systems:

* Rainwater harvesting captures and stores rainwater for non-potable use (not for drinking).

* Greywater reuse systems capture and store waste water from baths, washbasins, showers, sinks and washing machines.

Rainwater harvesting

In this system, water is harvested usually from the roof and then distributed to a tank. Here it is filtered and then pumped into the dwelling for reuse. The recycled water is usually stored in a cistern at the top of the building.

Greywater reuse

The idea of this system is to reduce mains water consumption. The greywater is piped from points of use, such as sinks and showers, through a filter and into a storage tank. The greywater is then pumped into a cistern where it can be used for toilet flushing or for watering the garden.

Waste management

The expectation within the construction industry is increasingly that working practices conserve energy and protect the environment. Everyone can play a part in this. For example, you can contribute at home by turning off hose pipes when you have finished using water.

Simple things, such as keeping construction sites neat and orderly, can go a long way to conserving energy and protecting the environment. A good way to remember this is Sort, Set, Shine, Standardise:

Sort – sort and store items in your work area, eliminate clutter and manage deliveries.

Set – everything should have its own place and be clearly marked and easy to access. In other words, be neat!

Shine – clean your work area and you will be able to see potential problems far more easily.

Standardise – using standardised working practices you can keep organised, clean and safe.

Reducing material wastage

Reducing waste is all about good working practice. By reducing wastage disposal and recycling materials on site, you will benefit from savings on raw materials and lower transportation costs.

Let's start by looking at ways to reduce waste when buying and storing materials:

* Only order the amount of materials you actually need.

* Arrange regular deliveries so you can reduce storage and material losses.

* Think about using recycled materials, as they may be cheaper.

* Is all the packaging absolutely necessary? Can you reduce the amount of packaging?

* Reject damaged or incomplete deliveries.

* Make sure that storage areas are safe, secure and weatherproof.

* Store liquids away from drains to prevent pollution.

By planning ahead and accurately measuring and cutting materials, you will be able to reduce wastage.

Statutory legislation for waste management

By law, all construction sites should be kept in good order and clean. A vital part of this is the proper disposal of waste, which can range from low risk waste, such as metals, plastics, wood and cardboard, to hazardous waste, for example asbestos, electrical and electronic equipment and refrigerants.

Waste is anything that is thrown away because it is no longer useful or needed. However, you cannot simply discard it, as some waste can be recycled or reused while other waste will affect health or the quality of the environment.

Legislation aims not only to prevent waste from going into landfill but also to encourage people to recycle. For example, under the Environmental Protection Act (1990), the building services industry has the following duty of care with regard to waste disposal:

* All waste for disposal can only be passed over to a licensed operator.

* Waste must be stored safely and securely.

* Waste should not cause environmental pollution.

The main legislation covering the disposal of waste is outlined in the table below.

DID YOU KNOW?

If waste is not properly managed and the duty of care is broken, then a fine of up to £5,000 may be issued.

Legislation	Brief explanation
Environmental Protection Act (1990)	Defines waste and waste offences
Environmental Protection (Duty of Care) Regulations (1991)	Places the responsibility for disposal on the producer of the waste
Hazardous Waste Regulations (2005)	Defines hazardous waste and regulates the safe management of hazardous waste
Waste Electrical and Electronic Equipment (WEEE) Regulations (2006)	Requires those who produce electrical and electronic waste to pay for its collection, treatment and recovery
Waste Regulations (2011)	Introduces a system for waste carrier registration

Table 3.15

Safe methods of waste disposal

In order to dispose of waste materials legally, you must use the right method.

* **Waste transfer notes** are required for every load of waste that is passed on or accepted.

* **Licensed waste disposal** is carried out by operators of landfill sites or those that store other people's waste, treat it, carry out recycling or are involved in the final disposal of waste.

* **Waste carriers' licences** are required by any company that transports waste, not just waste contractors or skip operators. For example, electricians or plumbers that carry construction and demolition waste would need to have this licence, as would anyone involved in construction or demolition.

* **Recycling** of materials such as wood, glass, soil, paper, board or scrap metal is dealt with at materials reclamation facilities. They sort the material, which is then sent to reprocessing plants so it can be reused.

* **Specialist disposal** is used for waste such as asbestos. There are authorised asbestos disposal sites that specialise in dealing with this kind of waste.

Recycling metals

Scrap metal is divided into two different types:

* **Ferrous** scrap includes iron and steel, mainly from beams, cars and household appliances.

* **Non-ferrous** scrap is all other types of metals, including aluminium, lead, copper, zinc and nickel.

Recycling businesses will collect and store metals and then transport them to **foundries**. The operators will have a licence, permit or consent to store, handle, transport and treat the metal.

Recycling plastics

Different types of plastic are used for different things, so they will need to be recycled separately. Licensed collectors will pass on the plastics to recycling businesses that will then remould the plastics.

Recycling wood and cardboard

Building sites will often generate a wide variety of different wood waste, such as off-cuts, shavings, chippings and sawdust.

Paper and cardboard waste can be passed on to an authorised waste carrier.

Disposing of asbestos

Asbestos should only be disposed of by specialist contractors. It needs to be double wrapped in approved packaging, with a hazard sign and asbestos code information visible. You should also dispose of any contaminated PPE in this way. The standard practice is to use a

KEY TERMS

Ferrous – metals that contain iron.

Non-ferrous – metals that do not contain any iron.

Foundry – a place where metal is melted and poured into moulds.

PRACTICAL TIP

Before collection, plastics should be stored on hard, waterproof surfaces, undercover and away from water courses.

PRACTICAL TIP

Sites must only pass waste on to an authorised waste carrier, and it is important to keep records of all transfers.

red inner bag with the asbestos warning and a clear outer bag with a carriage of dangerous goods (CDG) sign.

Asbestos waste should be carried in a sealed skip or in a separate compartment to other waste. It should be transported by a registered waste carrier and disposed of at a licensed site. Documentation relating to the disposal of asbestos must be kept for three years.

Disposing of electrical and electronic equipment

The Waste Electrical and Electronic Equipment (WEEE) Regulations were first introduced in the UK in 2006. They were based on EU law – the WEEE Directive of 2003.

Normally, the costs of electrical and electronic waste collection and disposal fall on either the contractor or the client. Disposal of items such as this are part of Site Waste Management Plans, which apply to all construction projects in England worth more than £300,000.

* For equipment purchased after August 2005, it is the responsibility of the producer to collect and treat the waste.

* For equipment purchased before August 2005 that is being replaced, it is the responsibility of the supplier of the equipment to collect and dispose of the waste.

* For equipment purchased before August 2005 that is not being replaced, it is the responsibility of either the contractor or client to dispose of the waste.

Disposing of refrigerants

Refrigerators, freezer cabinets, dehumidifiers and air conditioners contain **fluorinated gases**, known as chloro-fluoro-carbons (CFCs). CFCs have been linked with damage to the Earth's **ozone layer**, so production of most CFCs ceased in 1995.

Refrigerants such as these have to be collected by a registered waste company, which will de-gas the equipment. During the de-gassing process, the coolant is removed so that it does not leak into the atmosphere.

Key benefits of using sustainable materials

In summary:

* Using locally sourced materials not only cuts down on the transportation costs but also the pollution and energy used in transporting that material. At the same time their use provides employment for local suppliers.

* In choosing sustainable materials rather than materials that have to go through complex production processes or be shipped in from other parts of the world, construction should be more efficient and have a lower general impact on the environment.

* The use of energy saving materials will have a long-term and lasting impact on the use of energy for the duration of the property's life.

KEY TERMS

Fluorinated gases

– powerful greenhouse gases that contribute to global warming.

Ozone layer

– thin layer of gas high in the Earth's atmosphere.

* Not only will the construction industry have a lower carbon footprint, but also everything they build will have been constructed using lower carbon technologies and materials.

* Protecting the local natural environment from damage by construction work or surrounding infrastructure is only part of the environmental consideration. In choosing sustainable materials to use in construction projects the natural environment is protected elsewhere, by reducing quarrying, tree-felling and the use of scarce resources

* Recycling as much construction waste as possible, particularly from demolition, means that the industry will make less contribution to landfill. Most materials except those that are toxic or hazardous can be repurposed.

TEST YOURSELF

1. What area of the construction and built environment industry would be involved in examining planning applications regarding the long-term future of an area?

   a. Town planners

   b. Surveyors

   c. Civil engineers

   d. Construction site managers

2. What is the term used to describe transport routes such as roads, motorways, bridges and railways?

   a. Services

   b. Infrastructure

   c. Commercial

   d. Utilities

3. Which of the following is an example of a corporate client?

   a. Small business owner

   b. Local authority

   c. Government department

   d. Insurance company

4. What is another term that can be used to describe a land agent?

   a. Land surveyor

   b. Quantity surveyor

   c. Estate agent

   d. Building inspector

5. Which job role involves overseeing construction work on behalf of an architect or client to represent their interests on site?

   a. Clerk of works

   b. Main contractor

   c. Sub-contractor

   d. Building control inspector

6. Which of the following is an example of a renewable energy resource?

   a. Plants

   b. Sun

   c. Wind

   d. All of these

7. What does the National Green Specification Database provide?

   a. Methods on how to recycle

   b. A list of all recycling sites

   c. A list of environmentally friendly building materials

   d. A list of components required for building jobs

8. Which part of the Building Regulations focuses on energy conservation?

   a. Part B

   b. Part G

   c. Part H

   d. Part L

9. Which of the following is an example of biomass?

   a. Coal

   b. Peat

   c. Coke

   d. Logs

10. In addition to providing heating, which of the following also provides cooling?

    a. CCHP

    b. CHP

    c. MCHP

    d. HPCP

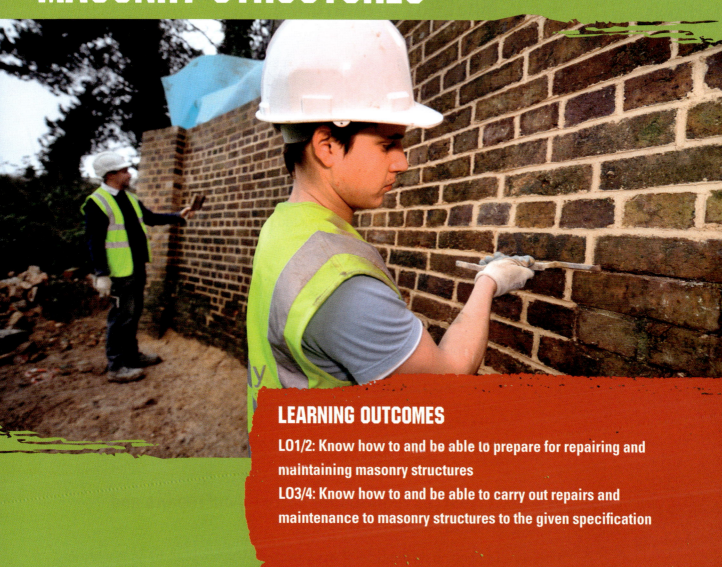

Chapter 4

Unit CSA–L3Occ132

REPAIR AND MAINTAIN MASONRY STRUCTURES

LEARNING OUTCOMES

LO1/2: Know how to and be able to prepare for repairing and maintaining masonry structures

LO3/4: Know how to and be able to carry out repairs and maintenance to masonry structures to the given specification

INTRODUCTION

The aim of this chapter is to:

* help you to select materials, components, tools and equipment

* help you to repair and maintain existing brick and block and vernacular style structures.

PREPARING TO REPAIR AND MAINTAIN MASONRY STRUCTURES

Although one of the major advantages of masonry construction is its durability, it is necessary to carry out inspections and maintenance to extend the life of the structure. In many cases the first time that repair and maintenance is planned for a building is when there is an obvious problem. This might involve problems with the damp course, the pointing in joints having fallen away, or damage done by plant growth or soil movement.

One of the biggest problems is always likely to be moisture penetration. This will affect a masonry structure no matter how well it has been designed or constructed. Neglecting maintenance will inevitably lead to further deterioration, which could in turn affect other parts of the building.

Health and safety and hazards

In Chapter 1 we looked at the importance of identifying potential hazards and the ways in which risk assessments and method statements can help to avoid possible health and safety hazards. When you are dealing with existing masonry structures it is often difficult to know precisely what the potential hazards will be. You will not know exactly how the building has been constructed or what hazards may lie hidden.

Resources required for carrying out repairs and maintenance

Depending on the type of work involved, you are likely to need a broad range of different resources. Some of these will have to be matched in order that they do not stand out as obvious repairs or replacements. This can be problematic, as some materials may have aged or may no longer be in production. The following table outlines some of the resources that you will probably need to obtain.

Resource	Explanation
Bricks	First of all you have to identify the type of brick that needs to be replaced. Is it a facing brick? If so, the match is going to have to be like for like. However, the existing bricks in the walling are likely to have aged and changed colour over time. You may have to consider sourcing bricks from a company that specialises in reclaimed materials. It is good practice to remove a brick that is representative of those needing replacement and to take it to the reclamation yard to try to match it.
Blocks	It is unlikely that blocks are going to be visible after they have been replaced. However, they need to be accurately measured and again like for like blocks obtained.
Natural stone	It is unlikely that two loads of natural stone, even if they were ordered at the same time and from the same source, are going to be identical. This makes the replacement of stone materials all the more difficult. You need to identify the type of stone used. Once again reclamation yards are a good place to begin the search for a suitable replacement.
Local materials	In parts of the country local materials are often used, making the buildings characteristically different. They may not be entirely made from local materials but may feature them as decorative parts of the building. Local building merchants and reclamation yards should be a good source of local stone or specific types of brick.
Mortars	The colour and texture of mortar can very much depend on the matrix and aggregate that was used, along with the time of year, incorporating the temperature and humidity. Each different bricklayer has their own way of mixing mortar and it can be difficult to achieve an exact match. The mortar used does need to be as close in colour to the original as possible and this may require a certain amount of experimentation to get the colour right.
Lightweight blocks	These should never be used to replace more solid structural blocks of other materials in a wall. Lightweight blocks are unlikely to be visible once installed, but dimensions and size are important, as is ensuring that they are not so wet that they will shrink later.
Insulating blocks	Insulation blocks are used for higher levels of thermal protection. They also conform to Part L of the Building Regulations.
Fixing devices	In some cases masonry walls may have been constructed without enough wall ties and anchors. These may have to be retro-fitted, usually by drilling holes in the masonry at a joint.
Window frames	Window frames need to be accurately measured and they also need to match the other existing ones in the building. It may be possible to buy matching frames brand new from suppliers or you may need to source them from reclamation yards.
Door frames	Replacing door frames presents the same kind of problems as window frames. They will need to both fit and match.
Wall extension profiles	These are wall connector systems that are designed to be used with either bricks or blocks. They are the way in which you can tie in new walls to existing masonry walls or columns.
Steel and concrete lintels	A lintel is a load-bearing component, such as a horizontal beam over an opening for a door or window. A structural engineer will calculate the type and size of lintel required and will specify whether it needs to be steel or concrete.
Horizontal brick reinforcement	These are metal bars or wires that are placed in horizontal bed joints to increase the strength of the masonry. They are used to help resist any stresses on a particular section of masonry work.
Cement	There is a broad range of different types of cement that can be used, each of which has its own particular features and suitability. Some are quick hardening, others are chosen for their particular resistance to damp, sulphates or frost.

Table 4.1 Resources for repairs and maintenance

Health hazards

In Chapter 3 we looked at some of the hazards associated with asbestos and how it needs to be handled. However, asbestos is not the only potential health hazard when carrying out repair and maintenance work:

* Dermatitis – this is an inflammation of the skin. It can be caused by exposure to wet cement, resins, hardeners, sealants, bitumen and solvents. It is a problem that bricklayers should certainly be aware of at all times. To reduce the risk of dermatitis you should always wear appropriate gloves and, if it is not possible to avoid handling these types of materials, you should always wash your hands and forearms and other exposed parts of the skin after exposure.

* Hand/arm vibration – if you are involved in breaking up concrete or using hammer drills, sanders, grinders or disc cutters then you could be prone to this. It is best known as vibration white finger and shows itself as a tingling and numbness to the fingers and the fingers going white. You should make sure you are using machinery that produces the least amount of vibration and that the machines are properly maintained and serviced. You should limit the amount of time you are exposed to the vibration and wear protective clothing to encourage good blood circulation.

* MSD (musculoskeletal disorder) – this is a term used to describe an injury that could affect your muscles, joints, tendons or spine. If you are carrying out repetitive tasks, handling relatively light loads or handling heavy objects then you could be at risk. This is true for bricklayers and block layers. Always try to use another method of moving heavier materials rather than lifting. Try to use lighter blocks if possible and reduce the carrying distance and the number of times you repeat the same body movements. Try to lay bricks at various heights throughout the day.

* Noise – this can cause temporary or permanent hearing loss. If you are using a power tool you could be at risk. Do not work for extended periods of time with noisy machinery. The machines need to be kept well maintained, with regular repairs carried out.

* Respiratory diseases – there is a high risk of breathing in hazardous substances. Anything that produces dust, even if it is invisible, can later cause problems such as asthma and breathing difficulties. Try to use water to keep dust down and wear a dust mask. If possible use a local exhaust ventilation system.

PRACTICAL TIP

Remember that if materials are wet they are heavier. They may also be more difficult to handle and more likely to slip.

PPE

In Chapter 1 we looked at personal protective equipment in some detail. But you should always ensure that your PPE includes the following:

* Safety goggles or safety glasses – these should always be worn to protect against eye injury.

* Gloves – these are essential if you are using machinery, moving heavy or rough objects or working in cold conditions.

* Safety footwear – these should always be worn to protect against sharp objects, uneven ground, objects falling on your feet and poor weather conditions.

* Hard hat – to protect your head from objects falling from above and from hitting it on structures.

* High visibility clothing (hi-vis) – these may be waistcoats, jackets, overalls, or clothing with reflective pads, to ensure you can be easily seen on site.

Checking resources

The two rules of thumb are:

* replace like with like when carrying out repair and maintenance work

* ensure that any material you use complies with current Building Regulations and guidance.

Calculations and formulae

In Chapter 2 we looked at resource requirements. The same calculations can be used to estimate what resources you will need in order to carry out repair and maintenance jobs. Many of these are simple numerical calculations that are used to work out perimeter, volume, mid-girth areas, wastage and bulking, quantities, costings and estimating reclaimed materials.

Preparing and mixing materials

It is important to look at exactly how you gauge and mix concrete and mortars either by hand or using a mixer. It is likely that you will be using a range of different hand tools, mixing machines and equipment.

Each different type of mix has a particular set of ingredients. Gauging means proportioning materials. The mixes are given three numbers. For concrete this might mean:

* 1 – use one part cement

* 2 – use two parts fine aggregate or sand

* 4 – use four parts coarse aggregate.

These are volumes or ratios – in this case 1:2:4.

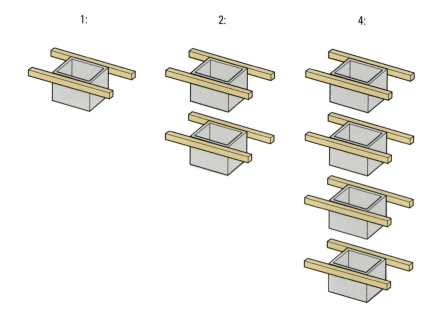

Figure 4.1 Mixing concrete by volume

Figure 4.2 One part of cement

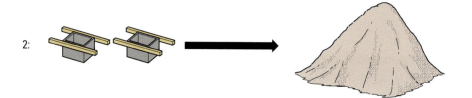

Figure 4.3 Two parts of fine sand

Figure 4.4 Four parts of coarse aggregate

Equipment used for mixing and how to use it

When you are mixing by hand you can either mix by volume or gauge the materials. If you are mixing by volume you can use the following equipment:

* A shovel – this is bad practice because it is a crude way of gauging the amount of material you are using. This is because the amount of material on a shovel can vary. There is more sand in a shovelful than there is if you have a shovelful of aggregate.

* A bucket – this is a better way to mix because if the bucket is full of the material it will have the same volume.

* Gauge box – this can be done using a wooden box, which is made to size. The boxes do not have bottoms and once you have gauged the material you can remove the box and shovel the material into the mix.

Another alternative is to use an increasingly popular method known as the dry silo system. A stand-alone silo is delivered with ready-mixed materials to the site. They are still dry. All that is required is that the correct amount of water is added to the silo to produce the required amount of wet material. This method is less likely to be used when undertaking repair and maintenance, however.

> **PRACTICAL TIP**
>
> Dry silo mortar systems, for example, hold around 33 tonnes of dry mortar. This is the equivalent of 23 cubic metres of mixed mortar.

Mixing construction materials using machines requires some additional equipment and materials. A good example would be mortar mixing, in which case you would need:

* the mixer – if it is petrol or diesel it will also need fuel; if it is electric then a waterproof cable connecting it to a generator or main supply will be needed

* gauge box

* wheelbarrows, buckets and shovels

* Portland cement

* plasticiser

* sand

* water.

All equipment and materials need to be placed close to the working area so that the mortar can be mixed and quickly moved to where it is needed. This also means having all of the ingredients for the mortar mix close to where the mixing is taking place.

CASE STUDY

Modern communications get the job done

Gary Kirsop, Head of Property Services at South Tyneside Homes, says:

'Technology is playing a massive part in our business these days. In responsive repairs and empty homes, we're striving to be 100 per cent paperless. We also want to provide a service that is highly responsive.

So, at 8am, our repairs team turn on their smartphone while they're still at home, log on to the system, and get the details of their first job … all before they've left the house. That means they don't waste time getting into work first, and they can go directly to the first job.

Each time they complete a job, they log the job and materials used, and then the details of the next job are downloaded straight to the phone. What used to happen is that they'd be given about 15 pieces of paper and they had to figure out what order to put all the jobs in themselves.

Working in this way has had a great effect on productivity because they waste less time on admin and get to the jobs more quickly. One of our employees said, "I used to be a van driver that occasionally did a bit of plumbing" – and now he's finally doing more plumbing than driving!'

Basic calculations

Calculating the exact mix you need to produce particular materials is going to depend on a number of things:

* the strength of the material you need

* the purpose of the material you need

* the specification for the job

* the hardening time of the material.

Gauging mortar

When you are producing mortar it is important that the mortar mix is the same throughout the whole job. You want a consistent strength and look. This often means that measuring the mix using shovels can be inaccurate.

The way around this is to get the proportions right by weight, although you can use gauge boxes and buckets to get the right proportion by volume.

If mortar is proportioned by volume you will see a ratio in the specification, as can be seen in the following table.

Mortar type and ratio	Actual ingredients
Cement mortar 1:6	One part cement and six parts sand
Lime mortar 1:6*	One part lime and six parts sand

Table 4.2 Ratios for mortar by volume

* Note that only specialists would mix lime mortar.

Examples of the ingredients needed for each of these types of mortar can be seen in the following diagrams.

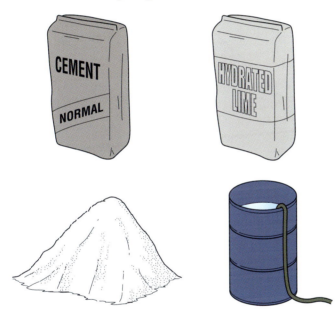

Figure 4.5 Cement/lime mortar – example of mix ratio

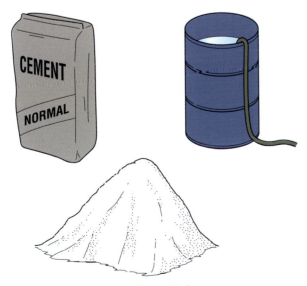

Figure 4.6 Cement mortar – example of mix ratio

Concrete

Concrete can be mixed either by hand or using a mixer. Again you will see a ratio mix, which, for example, might be 1:2:4. This means that for every one part of cement you will need two parts sand and four parts aggregate.

The following diagrams show how the process of mixing these ingredients works in practice.

Figure 4.7 Hand mixing concrete

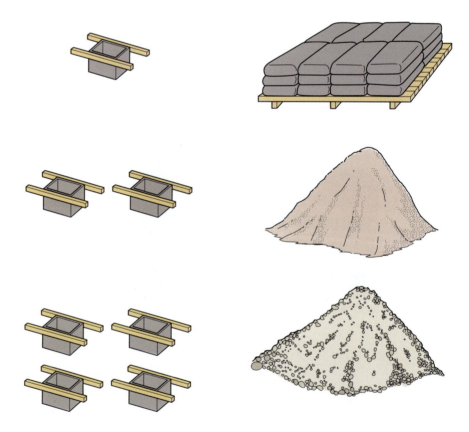

Figure 4.8 Materials for the concrete mix

Characteristics, uses and limitation of materials

There are often only four key ingredients to any particular mix:

* Mortar mixes – different strengths are required, depending on where the mortar will be used. 1:4 is the strongest mix and is used for engineering bricks, dense concrete blocks and work below the damp-proof course level. The higher the second number, the weaker the mix is, so normally a 1:6 mix will be used for bricks and blocks above the damp-proof course level.

* Sand – this has grains that are smaller than 5mm and are classed as a fine aggregate. The sand needs to be clean and free of any organic matter.

* Coarse cement aggregates – this means gravel or crushed rock. These have grains that are larger than 5mm.

* Water – always use potable (drinking) water and never use water from ponds, water tanks or any other source. Only drinkable water will be free from pollutants and organic matter.

Maximum time for use of mixed mortar, concrete and plaster

The maximum time to use mixed materials is very dependent on the weather conditions. In hot, drier weather the material is likely to have a shorter usage time.

Mortar tends to have a maximum usage time of around 2 hours.

Concrete is workable for around 1½ hours, but in hotter conditions this could be as little as 1 hour.

Plaster usually sets within 1½ hours but if it is cold then it can take as long as 2 hours.

Preparation times for hand-mixing

Mixing time is going to vary depending on the type of materials that you are using. The most important thing to ensure is that all of the particles that make up the mix produce a consistent wet material.

In a mixing machine, for example, mortar can be mixed within 2 to 3 minutes.

Each manufacturer will suggest their hand and machine mixing times. Multipurpose concrete, for example, which has graded sand and aggregate already included, will mix in a machine between 3 and 5 minutes. But this can take longer by hand, as you need to get the right workability and consistency.

Tools and equipment

The exact mix of tools and equipment that you are likely to need will depend on the type of job that you are doing. The following table outlines most of the common tools and equipment used on a regular basis.

Tools or equipment	Description
Hand tools	These will include the basic bricklayer's kit of bolster and hammer, trowels and other essential bricklaying, pointing and jointing tools and equipment.
Portable power tools	These will tend to be used when it is necessary to cut away sections of brick or blockwork, or when it is necessary to provide fixing points for other components. This would include cutters and drills as good examples.
Shoring equipment	This is necessary if you are required to dig trenches in order to access underground brick or blockwork (or foundations). These would include trench box systems.
Access equipment	This covers a very broad range of different ways in which you can gain access to work at higher levels. It would include ladders, hop-ups, stepladders, lightweight tower scaffolds, trestles and staging.
Mechanical aids for propping and supporting	Props and supports will often be necessary if excavation work is required, or if new openings are being created in existing masonry work. It may also be necessary to use steel props if major structural components such as lintels, or anything that has a high load-bearing function, has to be removed and replaced.

Table 4.3 Tools and equipment needed

CARRYING OUT REPAIRS AND MAINTENANCE TO MASONRY STRUCTURES

Maintaining standards

It is not sufficient to make a repair or perform the maintenance work and leave obvious signs that the work has been carried out.

If the repair work is being carried out to components of the structure that are not visible then this is not so much of a problem. It is when remedial work is required to clearly visible areas that the need to match and replace like with like becomes important. This may mean sourcing facing materials, such as bricks, with components sourced from reclamation yards. It may also mean having to provide a new surface over a broader part of the structure to cover both the repair and untouched parts of the structure.

Brickwork and stone comes in a huge variety of different shapes, colours, styles and textures. The materials that have come from a variety of different sources have their own characteristics. It is more important to look at the characteristics of brick and stone work on every individual building. The appearance of the brickwork will also be different according to the methods of bonding, decorative details and mortar used. It will sometimes be necessary to obtain matching stones and bricks from second-hand sources. These need to be checked in terms of colour, texture and size. It is important that they are also in good condition.

PRACTICAL TIP

Bricks that have been painted or plastered in the past can never be cleaned successfully and should not be reused. However, many brick manufacturers have a range of clay bricks, for example, and different decorative shapes. It is often possible to find good matches.

Replacement and repair of damp-proof barriers

Rising damp is a problem that can affect any building. It is a particular problem for older structures. The original damp-proof course may have become damaged, or perhaps the ground level around the property has changed as a result of new external features. Simply adding a flower bed or patio above the damp-proof barrier can cause rising damp.

All buildings are surrounded by natural moisture and materials such as brick are porous meaning that they will soak up moisture if they do not have the protection of a damp-proof membrane.

Rising damp can cause major problems in older properties. It can lead to severe decay in timbers. It is important to establish whether the damp-proof barrier is actually to blame. Looking at the building should establish whether the damp-proof barrier has been compromised by external features being added after the barrier was fitted. In this case, a solution might simply mean lowering external features, such as a drive, patio or flower bed, to stop the rising damp. The area will dry out naturally provided the affected rooms are heated and windows are left open for ventilation.

In some cases, however, the damp-proof barrier will have failed. There are methods that can be used in order to re-establish this barrier:

* It may mean fitting a new physical damp-proof membrane – this will also involve removing and replacing any affected plasterwork and damaged timbers.

* It may mean drilling holes into the affected area and injecting water-repellent chemicals into the brickwork. The chemicals are injected under pressure.

* In some cases it is possible to use injection mortar. This is a good solution where there is a good sized bed joint. The mortar, or cream, is injected under low pressure. The rising damp carries it up into the mortar bed. The mortar or cream crystallises and creates a new damp-proof barrier.

* A new treatment, known as electro osmotic, can be used to introduce an electric current into the wall just above ground level. This pushes the moisture back down the wall and dries out the area.

Temporarily supporting existing walls and floors

Certain jobs that involve structural operations may need temporary support. If any load-bearing elements, such as floors, roofs or walls, are going to be removed or altered then some kind of temporary support, or shoring, may be necessary.

Acrow props are telescopic tubular steel pieces of equipment. They can be adjusted using a large diameter screw thread. The screw thread allows for fairly fine adjustment. For larger adjustment and sizing there is a pin that slots into a series of holes in the inner tube of the prop. The props are made in a number of different standard sizes, from size 0 to size 4. This will give a broad range from 1 m to 5 m. The safe load for each of the props is clearly marked on it.

The acrow props can be used for a wide variety of different repair and maintenance work. Typically they will be used to support the stucture above a lintel or beam while you are replacing them. In these cases a pair of the props are used, one at either end of the opening.

What makes these pieces of equipment particularly useful is that the base and top plates are small so they can be fitted into fairly restricted areas. It is, of course, possible to lace or strut the props with scaffolding poles to stop them from falling over if necessary.

Figure 4.9 Acrow props lined up

Forming openings in existing masonry walling

The first task is to establish whether or not the masonry walling involved is loadbearing or non-loadbearing. New doorways of up to 2 m in width can be formed in non-loadbearing walls. Ideally these walls should not be more than 125 mm thick and 2.7 m in height.

The masonry above the new doorway should be supported by a reinforced concrete lintel. This needs to be bedded in mortar. The lintel should be 100 mm thick and 150 mm high. The lintel needs to have a seating of at least 150 mm on either side of the doorway. It also needs to be seated on a whole concrete pad stone.

The alternative is to put in a galvanised steel box lintel, which needs to be the full width of the wall. This needs to be bedded into the mortar and seated and supported in the same way as a reinforced lintel.

Care needs to be taken to make sure that any part of the wall being removed is not load-bearing. In all cases the location of a new opening has to be chosen with care. It is important to make sure that new openings do not add stress to any surrounding masonry. The consequence of this is that there will be cracking and distortion over time. The general rule of thumb is that there should be at least 600 mm of walling between any reveal and adjacent opening in any external wall.

Once the position has been worked out the following factors have to be taken into account:

* Sufficient permanent support in the form of a lintel will be required to hold up the wall above the opening.

* Calculations will be required to determine the size of the new lintel being added to support the opening. This should be ordered in advance; if it is a non-standard length, there may a lead time for manufacture and delivery.

* Check whether the opening is being formed in a cavity wall. The cavity will need to be closed and weather-proofed once the opening has been formed.

* The finishing and other features on either side of the wall affected should be removed or, if this is not possible, protected, before starting work.

* Depending on the height at which the new opening is to be formed, it may be necessary to use access equipment in order to reach the area safely. The type of access equipment used will depend on how much work at height is required and the surroundings you are working in.

* Temporary supports will have to be in place before, during and after the cutting of the opening until they are replaced with permanent support.

* Suitable portable power tools (such as angle grinders or disc cutters) will be necessary to cut through the existing masonry. Hammers and bolsters will also be required for the removal of brick and blockwork.

* There may be different floor levels on either side of the new opening. You will need to decide how this should be dealt with.

* Any damage to surrounding masonry, walls, floor and ceiling surfaces will have to be made good once work to create the opening is complete.

* Redecoration of the room may be required.

Replacing lintels

Even small changes to openings may require a new support beam or lintel. In some cases it is necessary to replace a lintel as a result of obvious cracking or damage to masonry. In these cases the existing lintel is probably not strong enough to cope with the stresses.

Lintels are used to bridge over openings. The weight carried by the lintel is transferred onto the piers either side of the opening. The lintels are the simplest way of carrying a load over an opening. The problem is that when lintels have their full load on them they have a tendency to bend. At some point the bend will reach a critical point where it will fracture. This means that the lintel needs to be right for the job in the first place. Unfortunately buildings can change their layout and function over time, and, in this case, older lintels are likely to suffer defects due to additional stresses. They may be at various stages of bending and failure.

In order to install a new lintel it is important to make an accurate measurement of the size of the structural opening. You should add 150 mm to each end to take account of the fact that the lintel needs to be firmly supported on either side.

Above the area of the lintel you will need to install needles and use acrow props. It is normal practice to use a hammer and chisel to remove the mortar around the bricks that are going to be removed. As soon as the mortar has been removed the bricks can then be lifted out of the walling to make room for the new lintel. The lintel is then laid on a fresh bed of mortar. It is important to allow the mortar to cure before any further work is carried out.

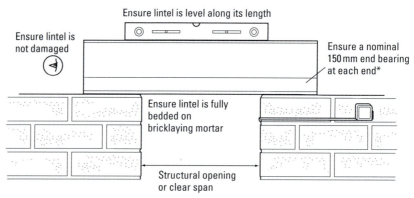

Figure 4.10 Lintel being installed

PRACTICAL TIP

You should carry out a risk assessment before starting any repair work. This should refer to:
- the method of work
- manufacturer's technical information
- statutory regulations
- official guidance.

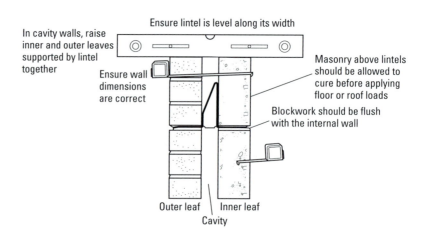

Figure 4.11 Cross-section of walling in preparation for bedding a lintel

PRACTICAL TASK

1. FORM AN OPENING AND FIT A LINTEL

OBJECTIVE

To cut out an opening and fit a lintel above it.

There will be occasions when a wall may require removing to make a bigger room or an opening cut into a wall to form a door or window.

Before you start knocking any walls down, you have to decide if it is loadbearing or not. A loadbearing wall is one that carries a load from above, e.g. floors or ceiling. In this case you will have to fit a lintel or RSJ (reinforced steel joist) to carry the load.

You need to check whether the wall supports the floor joists or continues up through the roof space and whether the floor joists run parallel or through the wall. This will determine the way you support the wall.

We will assume that the joists run parallel to the wall for this exercise.

Before you can start forming the opening, the floor above must be supported using props.

PPE

Ensure you select PPE appropriate to the job and site where you are working. Refer to the PPE section of Chapter 1.

TOOLS AND EQUIPMENT

4 adjustable props

Lump hammer and bolster chisel

Sledge hammer

Spirit level

4 scaffold boards

Hand held disc cutter

Figure 4.12 Supporting the wall

STEP 1 Place one of the scaffold boards on the floor, parallel to the wall. Leave sufficient room to work and don't forget that you will need some sort of working platform to reach the top of the wall.

PRACTICAL TIP

The scaffold boards should be long enough to go at least 1 m past the area of the openings.

STEP 2 Place a prop near the ends of the scaffold board, with the pins roughly in the correct position for the height required. You will need two persons to carry this out.

STEP 3 Put another scaffold board directly above the one on the floor, against the ceiling. Make sure that the props are plumb. Tighten the props until they support the scaffold board.

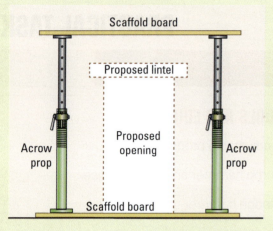

Figure 4.13 Placing adjustable props

PRACTICAL TIP

If you are taking out an area larger than a door opening, then you should place props no more than 1.5 m apart.

STEP 4 Fix nails through the holes in the plates on the props to stop them accidentally moving.

STEP 5 Repeat Steps 1 to 4 on the other side of the wall.

STEP 6 You are now ready to start cutting out the opening. Mark out the opening onto the wall.

PRACTICAL TIP

This can be done by either using chalk or a pencil to mark out the jamb or by fixing a piece of 25 × 50 mm timber up the sides of the opening to represent the jambs.

STEP 7 The method of cutting out the opening will depend on whether the premises are empty or occupied.

Method 1: Using disc cutter

If circumstances allow (for example, if the premises are empty), it would be quicker to use a disc cutter but this will be very dusty and you would need to take frequent breaks to get some fresh air.

Do not use a disc cutter if you have not been trained to use one.

Take care not to damage the surrounding area.

Method 2: Using hand tools

Always start at the top and remove any plaster to the marked area, including the lintel.

PRACTICAL TIP

It is sometimes a good idea to drill holes through the wall along the edges of the opening. This is time consuming but it makes it easier to cut the bricks/blocks.

Carefully cut out a hole at the top corner near the plumb line for the jamb. Ignore the lintel bearing points at this point.

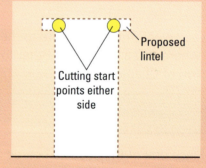

Figure 4.14 The starting position

Now continue to carefully cut out the rest of the lintel to the sides of the opening. If the mortar bed joint is soft then it may be possible to carefully cut that out first, followed by the bricks below.

STEP 8 Now continue down the sides of the opening. Whole bricks can be cut out course by course. Remove any debris at regular intervals; do not work with it under your feet – accidents can happen.

Carefully cut out the bearings. Do not damage the course below as this is where the lintel will sit.

Bed the concrete pad stone in place and allow it to set.

STEP 9 The lintel is now bedded into position. Do not put a mortar joint along the top of the lintel, but allow it to set. The top joint can now be wedged by using slate between the existing brickwork and the lintel, and then the joint should be pointed with mortar.

If this is done at the same time as bedding the lintel, the joint could shrink of the joint which could lead to subsequent movement at a later date.

STEP 10 Safely remove all the props and make good any damaged brickwork around the opening.

FORMING AN OPENING WHEN THE WALL IS SUPPORTING THE JOISTS

STEP 1 Once you have marked out the opening, cut out a hole above the lintel big enough to take the timber 'needle' – a piece of 150 × 100 mm timber.

STEP 2 Push the needle through the hole leaving an equal distance on both sides.

STEP 3 Set up two props, one on each side of the wall and adjust the height to support the needle. Then tighten the props to hold the needle. Check that the needle is level at this point and the props are plumb.

STEP 4 Complete cutting out the wall as described in Steps 7 to 10 above.

Repairing decorative features in existing brick walling

The existing type of brick bonding needs to be replicated when you are repairing decorative features in walling. These are important characteristics of the building and they need to be retained. It is always good practice to minimise the number of bricks that you have to replace. However you do need to make sure that you tackle any bricks or courses that have been damaged and extend the repair until you reach an area that is still sound.

The general rule of thumb is to try to conserve as much of the existing fabric of the wall as possible. Any repair work needs to employ the traditional materials and methods that were used when the building was originally constructed. Over the lifetime of the building the brickwork will have weathered, so size, colour and texture issues are very important. If existing bricks cannot be reused then carefully selected second-hand bricks are a good alternative. There are numerous reclamation companies that specialise in this type of supply. There are also newly made bricks that aim to replicate particular types of traditional brick.

It may be necessary to clean the wall surface, although sand and grit blasting is not generally a good idea. A mix of water and acid brushed on will remove paint from brickwork without damaging the bricks.

PRACTICAL TIP

It can be very difficult to replace like for like but minor differences can often be hidden by ensuring that the jointing and pointing methods are correct and that a good match has been made when choosing the replacement bricks.

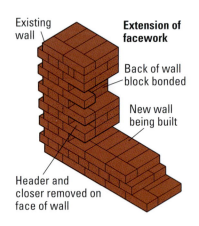

Existing wall

Extension of facework

Back of wall block bonded

New wall being built

Header and closer removed on face of wall

Figure 4.15 Cutting out of headers and stretchers for indents

Providing for walling extension

It is possible to lengthen a wall and still maintain the general face bond. This means cutting out recesses or indents on every alternate course. In other words, this means cutting out the header and closer. This can be seen in Fig 4.15.

Cutting out the header and closer has to be done with great care. It is very easy to damage the surrounding brickwork. This means that it is a time consuming job.

It is also common to use what is known as a starter tie, or wall starter system. The idea is that these ties will join new walls to existing masonry. There is a screw-in tie with a nylon plug, which can be seen in Figs 4.16 and 4.17.

In effect both of these systems (shown in Figs 4.16 and 4.17) give stability to the new wall at its start point from the existing wall. The tie is embedded into the existing wall and becomes part of the new structure of the extension wall.

Step-by-step guides to achieving walling extensions to existing walls can be seen in the practical exercises at the end of this section.

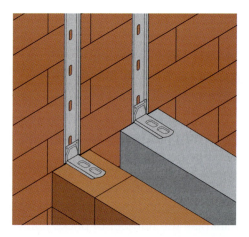

Figure 4.16 Universal wall starter system

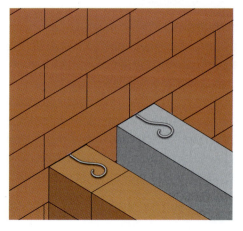

Figure 4.17 Starter tie

Jointing and pointing methods

It is essential that any repair work on decorative features in brick walling uses the right mortar. It is probable that a lime mortar, which is a 1:2:9 cement/lime/sand mix will be correct though this is mainly used in the heritage sector these days. More modern cement mortars are generally not a good match.

Repointing is a process of repairing defective mortar joints between the bricks that have worn away due to the weather or have cracked due to movement of the wall.

It is possible to point either as the work proceeds on the repair or after the brickwork has been completed. If you are pointing as the work proceeds then a few bricks are laid and then the joints are pointed. This is generally known as jointing. The following table shows the main advantages and disadvantages of jointing.

Advantages	Disadvantages
The face joints and the mortars within the walling are unified. This will minimise any possibility that the pointing can be pulled out by frost.	It is difficult to maintain a uniform colour in the mortar, so you need to keep a watch on the sand that is being used and how the mortar is being mixed.
This is a cheaper and quicker method of pointing because the wall is being completed as you are repairing it.	It is difficult to use a contrasting coloured pointing mortar.
It is not necessary to rake out the joints or refill them.	If the conditions are wet the joints are difficult to point and there is a chance that the repair will not be successful.

Table 4.4 Pointing and jointing

The other alternative is to point after raking out the joints. This has an effect of making the joints appear wider than they actually are, so it is a difficult thing to do if you are working on decorative features. It is better to use a sharpened piece of wood to do the raking out. The piece of wood should be cut to the width of the joint.

Once the repair has been completed it is important to clean the area down with water. You should use a scrubbing brush rather than a wire brush, which is only good if you are using hard facing bricks.

Once you have cleaned the repair area down you must allow it to dry out for the following reasons:

* It is difficult to point a wall that is saturated because the face of the bricks may become soiled with pointing mortar.

* The more water there is the more shrinkage movement there is likely to be when it dries out. This will break the bond between the bed joints and the pointing mortar. It will allow moisture to become trapped, which will freeze in poor conditions, making it expand and creating bigger cracks.

However, the wall should be a little damp, as this encourages adhesion and keeps shrinkage movement to a minimum.

The specific type of pointing needs to be identified. The following table describes some examples of typical types of pointing.

Type of pointing	Description
Flush 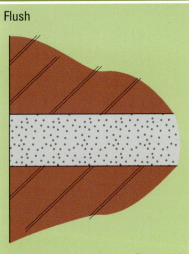 Figure 4.18	The mortar is pressed into the joint and finished flush with the brick's surface. This is common when hand-made facing bricks have been used.
Struck 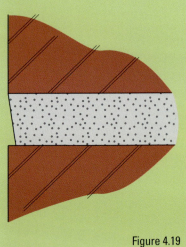 Figure 4.19	This is not usually used externally because water will collect on the upper part of the brick, which could cause damage when it freezes.
Weather struck and cut  Figure 4.20	This is a good all-round type of pointing for external work. The bed joints are pressed back on the upper edge of the joint. One edge of the perpends is cut and also the lower edge of the bed joints. This means that when bricks that are of an uneven size have been used the joints can still be made to appear equal in width and appear straight. It provides a very neat and uniform appearance.

Type of pointing	Description
Rounded or tooled 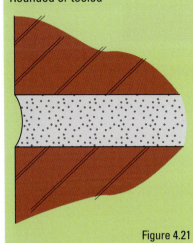 Figure 4.21	This is also known as bucket handle pointing. The joints are hollow as a result of rubbing a jointing iron over them. This means that the pointing has been pressed well into the joint, making it dense. It has a good effect when the bricks and joints are of regular shape and size.
Recessed 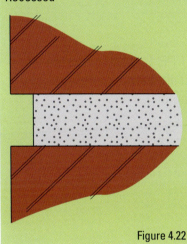 Figure 4.22	This is not usually recommended for external face work unless the bricks are hard and dense. The recess is formed by a brick jointer that is the same width as the joint. It tends to be used for internal face brickwork or on brick fireplaces. It works well when the bricks are of a regular shape and size.

Table 4.5 Some typical types of pointing

CASE STUDY

South Tyneside Homes

South Tyneside Council's Housing Company

Mobile working really works

Dave Melia, a Property Services Manager at South Tyneside Homes says:

'Mobile working might sound like a strict regime, but it brought a level of trust with it as well. They go out with a van that's fully stocked, so we're trusting them to do any extra little jobs at the same house that needed doing, rather than having to go back there another time. Another bonus of working like this is that it gives some flexibility to their diaries, for example, it means that a job can't be put in the system for them if there are conflicts with other commitments, e.g. picking the kids up from school.

We do have an ageing workforce and the older employees have had to learn the new technologies as well, but the younger ones are really helpful, showing them how it all works. It's nice to see the team working together like this.'

2. CUT OUT TOOTHING TO EXTEND A WALL

OBJECTIVE

To extend a wall by cutting out toothing.

There may be times when existing walls may have to be altered or extended. This can be done in a variety of ways, including:

* toothing

* block indents.

Toothing is when bricks are cut out at the end of the wall to be extended. The bricks are cut out in a way that does not affect the bond of the wall.

Care must be taken when cutting out toothing. Always start from the top and work down as this helps to prevent the projecting brick from snapping off.

The perp joints must be kept plumb to ensure that the bond is maintained along the extension.

Block indents are used to tie a new wall to an existing wall where the appearance is not important. They are similar to toothing as they are cut out of the wall to allow you to extend the wall but consist of cutting out three courses of brickwork instead of one, then leaving three courses.

Block indent   Toothing

Figure 4.23 Methods of extending a wall

STEP 1 Mark out the toothing that you are going to cut out.

TOOLS AND EQUIPMENT

Walling trowel

Pointing trowel

Spirit level

Lump hammer and bolster chisel/plugging chisel

Tape measure

PPE

Ensure you select PPE appropriate to the job and site where you are working. Refer to the PPE section of Chapter 1.

STEP 2 Starting at the top of the wall, cut out the joints around the brick that you want to remove.

Figure 4.24 Cutting out the joints around the brick to be removed

STEP 3 Continue until all the toothing is cut out.

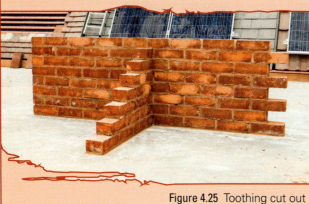

Figure 4.25 Toothing cut out

PRACTICAL TIP

At this point it is a good idea to check that all the toothings are clean, otherwise when you are building the extension you will not fit the bricks into the toothing.

STEP 4 Set out the wall extension and make sure that the bond is correct where the two walls meet.

STEP 5 Bed the first brick in position at the end of the wall. Make sure that it is level with the existing wall before you lay the first course.

Figure 4.26 Placing a brick into the toothing

STEP 6 Once you have laid the first course, and are happy with the bonding arrangement where the two walls meet, build a small corner.

STEP 7 Start running in the wall between the new corner and the existing wall.

PRACTICAL TIP

When you come to the toothing you must ensure that all the joints are full. If not, it will result in a weakness at this point and could lead to shrinkage cracking. Also make sure that the brick is level when you place it into the toothing.

STEP 8 Continue to build the wall in the same manner until the wall is complete.

Figure 4.27 The completed wall

PRACTICAL TIP

Block indents are carried out in the same way except that you cut out blocks of three courses instead of one.

PRACTICAL TASK

3. REPOINT DEFECTIVE JOINTS

OBJECTIVE

To practise repointing defective joints.

PPE

Ensure you select PPE appropriate to the job and site where you are working. Refer to the PPE section of Chapter 1.

TOOLS AND EQUIPMENT

Walling trowel

Pointing trowel

Lump hammer and bolster/plugging chisel

Hand hawk

Disc cutter/angle grinder

Jointing iron

Brush

STEP 1 Starting at the top of the wall, rake out the joints. There are several options:

1. Lump hammer and plugging chisel – this is very time consuming.

2. Disc cutter – you have to be very careful not to damage the bricks when cutting out the mortar joint.

3. Angle grinder and mortar raking tool – this is the better method if you are using mechanical tools. The raking tool attaches to the angle grinder and you can safely cut out the joint to the correct depth (15 mm) without damaging the bricks.

STEP 2 Dust down the area that you have just raked out with a soft brush to remove any surplus material.

STEP 3 Now damp down the area to reduce the chances of the mortar drying out too quickly. This would cause the mortar to set too quickly and become crumbly.

STEP 4 Starting with the perp joints, then the bed joints, re-point the area. Make sure that you press the mortar firmly into the joint.

Joint raked out

Joint re-pointed

Figure 4.28 Joints raked out correctly

STEP 5 Once the mortar has hardened you can go over it with a soft brush to take off any surplus mortar.

PRACTICAL TIP

The mortar should be as near as possible to the existing mortar in the wall. If you make the mortar too strong or too weak then it will cause problems in the future due to movement and the weather and will have to be re-pointed again.

REED TIP

If you're having trouble completing a particular task and find yourself getting frustrated, take a step back, work on something else that needs doing, try to approach it from a different angle, or use a fresh set of eyes – someone else might be able to see where you're going wrong.

PRACTICAL TASK

4. REMOVE DEFECTIVE BRICKS

OBJECTIVE

To practise removing and replacing bricks.

Sometimes you need to remove defective bricks from a wall. The damage could have been done in several ways but the most common one is the face of the brick breaking off (spalling) due to water penetration or frost damage.

Take care not to damage the bricks that are going to be left in, otherwise you will have to remove them too.

TOOLS AND EQUIPMENT

Walling trowel

Pointing trowel

Lump hammer and bolster/plugging chisel

Disc cutter

PPE

Ensure you select PPE appropriate to the job and site where you are working. Refer to the PPE section of Chapter 1.

STEP 1 Identify the defective brick/s and mark them with chalk.

STEP 2 Starting with the joint along the top of each brick, carefully cut out the bed joint with the hammer and plugging chisel.

STEP 3 Continue round the brick until all the mortar joint is cut out.

STEP 4 Starting in the middle of the brick, start breaking the brick up until you have cut it all out.

STEP 5 Using the same coloured brick and mortar, bed the brick into the hole. Leave the top joint until the other brick joints have hardened sufficiently.

PRACTICAL TIP

It is easier to use a drill and to drill as many holes into the brick as possible. This weakens the brick and makes it easier to cut the brick out.

Another method is to use a joint raking attachment for an angle grinder. This tool screws onto the grinder and cuts through the mortar joint. You can then move around the brick removing all the mortar and the brick should slide out easily.

STEP 6 Once the mortar joints have hardened sufficiently, you can 'pin up' the top of the brick by tapping in some slate between the bricks. This is done to prevent any shrinkage cracks appearing at a later date.

TEST YOURSELF

1. To which part of the Building Regulations does an insulating block need to conform?

   a. Part B

   b. Part D

   c. Part H

   d. Part L

2. What is toothing?

   a. A type of bond

   b. When bricks are cut out at the end of a wall that is going to be extended

   c. A method of raking out a joint

   d. Another name for block indents, when you cut out blocks of three courses

3. What size grains do coarse aggregates have?

   a. Less than 5mm

   b. More than 5mm

   c. More than 10mm

   d. More than 12mm

4. Which of the following can cause dermatitis?

   a. Sealants

   b. Solvents

   c. Wet cement

   d. All of these

5. What is the usual working time for mortar?

   a. Less than 1 hour

   b. Less than 1.5 hours

   c. About 2 hours

   d. About 3 hours

6. When injection mortar or cream is used as a damp-proof barrier, what happens to it when it reaches the mortar bed?

   a. It spreads into the mortar

   b. It crystallises

   c. It turns into a liquid

   d. It spreads into the brick

7. In millimetres, how much should you add to each side of an opening to ensure that a lintel is firmly supported?

   a. 50

   b. 100

   c. 150

   d. 200

8. What type of pointing can be described as having the mortar pressed into the joint so that it is at the same level as the brick surface?

   a. Weather struck

   b. Rounded

   c. Recessed

   d. Flush

9. Which of the following statements is TRUE when you are about to start pointing?

   a. The walls should be saturated with water

   b. The walls should be absolutely dry

   c. The walls should be damp

   d. Only the joints should be saturated with water

10. What is the minimum gap in millimetres that should be left between any existing reveal or opening and a new opening in a wall?

    a. 600

    b. 1,200

    c. 1,500

    d. 1,800

Chapter 5

Unit CSA–L3Occ131
CONSTRUCT CHIMNEYS AND FIREPLACE STRUCTURES

LEARNING OUTCOMES

LO1/2: Know how to and be able to interpret information to plan for constructing chimneys and fireplace structures

LO3/4: Know how to and be able to construct fireplaces and flues to the given specification

LO5/6: Know how to and be able to build decorative chimney stacks to the given specification

INTRODUCTION

The aims of this chapter are to:

* help you to select materials, components, tools and equipment

* help you to construct decorative brickwork fireplaces, flues and chimney stacks.

INTERPRETING INFORMATION TO PLAN FOR CONSTRUCTING CHIMNEYS AND FIREPLACE STRUCTURES

Figure 5.1 A modern chimney

Building Regulations Part J sets out to ensure that appliances producing heat are designed and installed to work efficiently and safely without giving rise to nuisances and without harmful effect on the structure. Adequate provision has to be made for the discharge of the products of combustion from the heating appliances to the outside air. Heat producing appliances, flue pipes and chimneys are to be installed and/or constructed so the risk of the building catching fire from these sources is reduced to a reasonable level.

Health and safety and hazards

Many of the key health and safety hazards associated with the construction of chimneys and fireplaces are common to those encountered in most brickwork jobs. You should refer to Chapter 1, which covers issues such as working at height and the need to wear appropriate PPE, which your employer should provide.

Many of the potential hazards can be identified by referring to a work method statement. This is a safe system of work and is produced after a risk assessment has taken place. It aims to:

* detail how the work has to be completed

* outline hazards

* provide a step-by-step guide on how to carry out the work

* include control measures to ensure safety.

The main point behind a work method statement is to comply with health and safety legislation. It is particularly important when working at height, as may be necessary when constructing chimneys. Often businesses will have to provide these statements as part of the process of getting a job.

A comprehensive work method statement should have four sections, which are:

* basic information about the company, standard operating procedures and a brief description of the work

* a summary of the main hazards, control measures and any PPE required

* a fuller description of the work and general information about staff, training, welfare, first aid, etc.

* a step-by-step to carrying out the task in a safe way.

Checking drawings, specifications and schedules

You should always check drawings, specifications and schedules and ensure that you use measurements from the site drawings.

Specifications are closely linked to working drawings. They provide additional information that is not usually on the working drawings. They could highlight any particular problems, detail the standard of work required including what materials are to be used, and cover other issues, such as working hours on site or whether the site has limited access. A version of a specification form can be seen in Fig 5.2.

Other features that can be found on the specification form may relate to the quality of materials that are to be used.

QUALITY SPECIFICATION		
Operation	**Standards**	**Comments**

Figure 5.2 Specification form

A schedule can be either prepared by the builders themselves or designers on a larger job. The schedule is also used alongside working drawings. They are useful for the following reasons:

* They detail the quantities of any materials that will be needed.

* They detail the physical characteristics of materials, such as type or size.

* They indicate where the materials will be placed on the site.

REED TIP

A good way to support your team is to avoid being late or taking unnecessary sick leave. Of course, everyone should be allowed to take sick leave if they are unwell, but remember that when you are away from work, it affects your workmates and your customers.

Contract documents

A great deal of construction work is carried out using a Standard Form of Contract such as those produced by the Joint Contractors (JCT) or the Buildings Employers' Confederation (BEF). The actual contract may well depend on:

* the type of client (whether this is a private individual or a company or perhaps a local authority)

* the size and type of the project (and whether there is a need to have sub-contractors working on part of the job)

* the type of contract documents attached to the main contract (such as a Bill of Quantities).

A Bill of Quantities will show:

* the architect's working drawings

* the specification for the job

* the schedule for the job.

These types of document tend to be used for larger projects. They are prepared by the architect, the quantity surveyor or the contractor themselves.

Interpreting measurements from drawings

Drawings are covered in detail in Chapter 2. For brickwork, drawings can be scaled using one of the following:

* 1:5
* 1:10
* 1:20
* 1:50

Construction or working drawings and sketches are essential. These are diagrams that show both the building specifications and the procedures. In most cases the architect will tend to use computer aided design packages or may draw the diagrams themselves. Sketches are simply that – they describe the particular requirements of either a construction procedure or a component.

Reporting inaccuracies

Occasionally there may be mistakes on working drawings or other documents. These can affect the way in which you will have to carry out the work. Typically, the following problems could occur:

* The scale and therefore the measurements on the drawings may be inaccurate once you have begun to measure out and set out the work.

* Vital information may be missing. A document may not have been provided, or information may be missing from one of the drawings or documents.

* The information on one of the drawings may be different from another document, such as the specification.

In all cases it is sensible to refer back to the individual who prepared the working drawings or documents and ensure that the errors are not carried through to the construction work.

You need to refer to the drawing number and date to make sure you have the latest drawing and/or specification when making reference back to the originator to clarify any details.

CONSTRUCTING FIREPLACES AND FLUES

Any heating method (including appliances) that will be used in a fireplace according to the Building Regulations must:

* have an adequate air supply for combustion and for the efficient working of a flueliner or chimney

* discharge the products of combustion to the outside

* reduce the risk of the building catching fire.

Chimneys, flues and hearths must be constructed to ensure the following:

* The materials used are of a non-combustible nature and of such a quality and thickness that they will not be adversely affected by heat, condensation or the products of combustion.

* They will be of such thickness as to prevent ignition of any part of the structure.

* They will prevent any smoke or products of combustion escaping into the building.

All flueliners must be:

* placed or shielded so as to ensure that, whether the liner is inside or outside the building, there is neither undue risk of accidental damage to the liner nor undue danger to persons in or about the building

* properly supported

* discharging into a chimney or open air.

If provision is made for a solid fuel fire to burn directly on a hearth, secure means of anchorage for an effective fireguard must be provided in the adjoining structure. If a flue serves an appliance which burns solid fuel or oil, the opening into the flue must be constructed so as to enable the flue to be cleaned and must be fitted with a closely fitting cover of non-combustible material.

Provision for services in fireplace construction

Before any installation of services, in this case heating appliances, it is important to look at the Building Regulations Part J. These give detailed advice on this type of work. There is also a Building Control consent requirement unless the work is being carried out by an individual who is registered with a 'competent person's scheme'.

Each different type of service has its own list of registered members:

* the Heating Equipment Testing and Approval Scheme (HETAS) for solid fuel and wood burning appliances

* the Gas Safe Register for gas-burning appliances

* the Oil Firing Technical Association for the Petroleum Industry (OFTEC) for oil-fired appliances

* the National Association of Chimney Sweeps (NACS)

* the National Association of Chimney Engineers (NACE).

Damp-proof barriers

Older chimneys were often constructed without a horizontal damp-proof membrane or with anything to serve that purpose above the roof line. This meant that damp could enter the chimney stack and soak down into the building. These types of chimneys can also suffer from rising damp as they attract moisture through condensation (they are a cold bridge). In time this can mean trouble for roof timbers.

Chimney stacks that are built from brick, block or stone can have a tendency to develop porous pointing and rendering as they get older. The levels of damp in older chimneys can be reduced, but never really eliminated.

Maintaining industrial standards

It is most important to ensure that no materials are built around any fireplace recesses in such a position that they are liable to create a fire hazard. The Building Regulations lay down the following minimum requirements:

- No combustible material should be placed in a chimney or fireplace recess, so as to be nearer to the flue, or the inner surface of the recess, than 150 mm in the case of a wooden plug or 200 mm in the case of any other material.

- Where the thickness of solid non-combustible material surrounding a flue is less than 200 mm, no combustible material other than a floorboard, skirting board, dado rail, picture rail, mantel shelf, or architrave should be placed nearer than 38 mm.

- No metal fastening which is in contact with combustible material should be placed in any chimney or fireplace nearer than 50 mm to a flue or the inner surface of a fireplace recess.

Forming openings, bridging fireplace openings and forming flues

The opening area of a fireplace should be calculated from the following formula:

Fireplace opening area (mm²) = Total horizontal length of fireplace opening L (mm) × Height of fireplace opening H (mm)

The fireplace opening should be:

- no greater than eight times the area of the flue (if the chimney is square or rectangular)

- no greater than ten times the area of the liner (if you are fitting a round metal chimney liner).

Fireplace recesses should be constructed in the following way:

- The jamb on each side of the opening is not less than 200 mm thick.

- The back of the recess is a solid wall not less than 200 mm thick or a cavity wall, each leaf of which is not less than 100 mm thick; such thickness must extend for the full height of the recess. If the recess is situated in an external wall and there is no combustible cladding across the back of the recess then this back may be a solid wall less than 200 mm thick but not less than 100 mm thick. If the recess serves as the back of each of two recesses other than in a separating wall, then it may be a solid wall less than 200 mm thick but not less than 100 mm thick (Fig 5.3).

- There is no opening in the back of a fireplace recess which does not communicate with a flue.

- No combustible material, other than timber fillets supporting the sides of the hearth where it joins the floor, may be placed under a constructional hearth serving as appliance within a distance of 250 mm measured vertically, unless this material is separated from the underside of the hearth by an air space of not less than 50 mm.

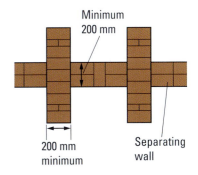

Minimum 200 mm

200 mm minimum

Separating wall

Figure 5.3 Plan showing the maximum thickness of the back of a fireplace

* An ash pit may be constructed under a constructional hearth if the sides and bottom are constructed of non-combustible material and not less than 50 mm thick; no combustible material may be built into a wall within 225 mm of the inner surface of the pit. Any combustible material placed elsewhere must be separated from the outer surface of the pit by an air space of at least 50 mm.

* If a duct is to be constructed under a hearth for the admission of combustion air to an appliance then the duct must be smoke-tight and constructed of non-combustible material.

Building fireplaces and flues

The floor of a fireplace is called a hearth. Hearths in ground floors may be supported by fender walls (Fig 5.4). If the hollow space under the hearth and behind the fender walls is filled in solidly with clean hardcore, there is no need to use any reinforcement in the concrete hearth.

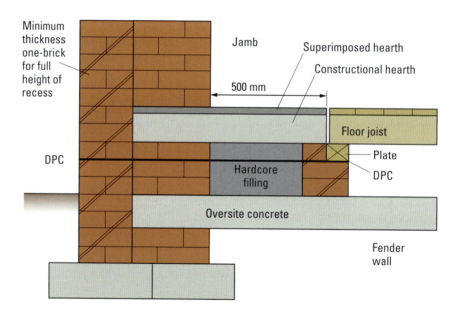

Figure 5.4 Section through a ground floor fireplace recess

With appliances, it is also necessary to provide a hearth in order to reduce the fire risk. Each appliance must have a constructional hearth which must be:

* not less than 125 mm thick

* not lower than the surface of any floor built of combustible material (Fig 5.5).

* extended within the recess to the back and jambs of the recess and projected not less than 500 mm in front of the jambs and not less than 150 mm each side of the jambs (Fig 5.5)

* not less than 840 mm square if the hearth is not constructed within a recess (Fig 5.6).

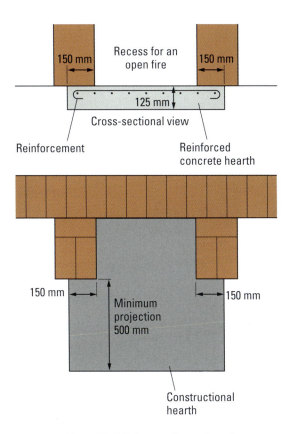

Figure 5.5 Minimum dimensions for a constructional hearth

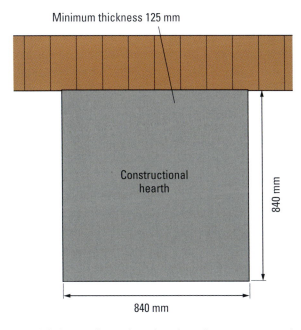

Figure 5.6 Minimum dimensions for a hearth not constructed within a recess

The simplest method of forming a hearth in an upper floor is to use reinforced concrete. If such a hearth is being supported by the wall, it is most important to remember that it is a cantilevered slab and therefore the reinforcement must be placed within the top part of the slab, about 25 mm from the upper surface. If timber formwork is used between the trimmer joist and the fireplace recess so that the concrete hearth may be cast *in situ,* such timbering must be removed if it is within 250 mm of the upper surface of the hearth (Fig 5.7).

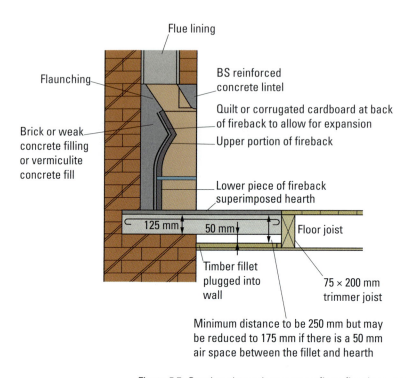

Figure 5.7 Section through an upper floor fireplace recess

A flue should be at least 175 mm in diameter (Fig 5.8) and must be lined to provide a chimney which is resistant to acids and the products of combustion, particularly in the case of solid fuel burning appliances. The three main substances obtained from the burning of coal are water, sulphur dioxide and carbon dioxide. When either of the two dioxides is mixed with the water they will produce acids: sulphurous, in the case of sulphur dioxide and carbonic in the case of carbon dioxide.

Although these acids are not in a concentrated form, their continual application inside a flue will eventually harm the mortar and, in some cases, the bricks. When the gases leave the fire they are hot and contain a lot of moisture vapour (where the temperature of the gases is above 100°C this would be in the form of steam), but as it cools down it is capable of holding less moisture in suspension. If this happens inside the chimney, the flue gases will condense into acid on the sides of the flue, and it is this that is harmful to the construction of the chimney.

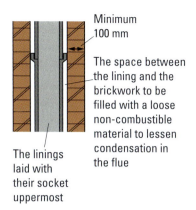

Minimum 100 mm

The space between the lining and the brickwork to be filled with a loose non-combustible material to lessen condensation in the flue

The linings laid with their socket uppermost

Figure 5.8 Section showing the method of building in lining and insulating them against excessive condensation

With slow burning appliances, the flue gases do not flow so rapidly, so there is a greater risk of these gases condensing much earlier than with open fires. It is, therefore, a very good practice to insulate the lining of the flue with a loose non-combustible material, for example, exfoliated mica or mineral wool. This will allow the lining to warm up quickly because it will not lose its heat to the surrounding brickwork too rapidly. Thus, the flue gases will not cool off so quickly and the amount of condensation will be kept to a minimum.

The more efficient an appliance is, the more products of combustion are obtained from the fuel, which means, in turn, that chimneys and flues must be much more resistant to acids.

If a chimney is constructed of bricks or blocks of concrete and lined with one of the materials previously described then any flue in the chimney must be surrounded and separated from any other flue in the chimney by solid material not less than 100 mm thick, excluding the thickness of the lining.

If a chimney is situated in a separating wall then the part of the chimney forming part of this separating wall must be not less than 200 mm thick (Fig 5.9(b)). In a cavity wall, each leaf should not be less than 100 mm thick (Fig 5.9(a)). A flue in a chimney serving an appliance must be such as will contain a circle having a diameter of not less than 175 mm measured in cross-section (Fig 5.9(b)).

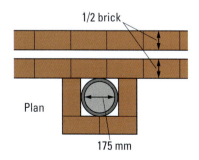

1/2 brick

Plan

175 mm

(a) The formation of a flue in a cavity separating wall

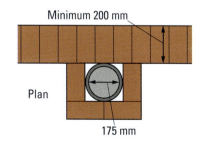

Minimum 200 mm

Plan

175 mm

(b) The formation of a flue in a one-brick separating wall

Figure 5.9 Flue formations

If a flue does not link to a fireplace recess, it must terminate at its lower end in a chamber which:

- has a means of access for inspection and cleaning fitted with a close-fitting cover

- is capable of containing a condensate collecting vessel.

No flue should make an angle with the horizontal of less than 45° and no flueliner should pass through any roof space, internal wall or partition.

The outlet of any flue in a chimney must be so situated that the top of this chimney (excluding any chimney pot) is not less than 1 m above the highest point of contact between the chimney and the roof (Fig 5.10).

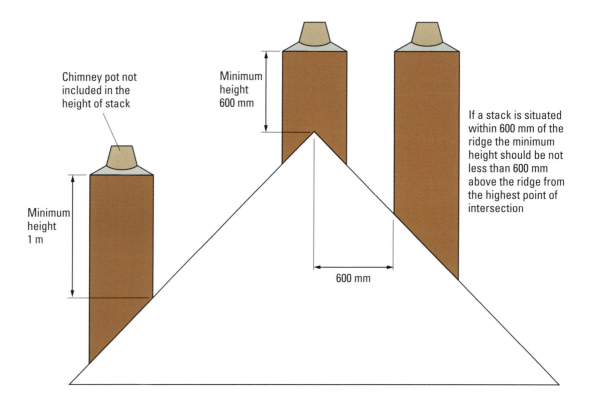

Chimney pot not included in the height of stack

Minimum height 600 mm

If a stack is situated within 600 mm of the ridge the minimum height should be not less than 600 mm above the ridge from the highest point of intersection

Minimum height 1 m

600 mm

Figure 5.10 The minimum heights of chimney stacks at different positions on the roof

If the roof has a pitch on each side of the ridge of not less than 10° and the chimney passes through the roof within 600 mm of the ridge, the top of the chimney should be not less than 600 mm above the ridge (Fig 5.10).

It should not be less than 1 m above the top of a window or opening skylight, which is not more than 2.3 m measured horizontally from the chimney.

Flue blocks and linings

Any linings or blocks must be jointed and pointed with cement mortar and any linings should be so built into the chimney so that the socket of each component is uppermost (Fig 5.8).

They should be made from fireclay or terracotta, be free from defects and give a clear ring when struck with a light hammer. They may or may not be vitrified; or be either glazed or unglazed.

Clay flue linings are available in straight or curved units. Straight units are nominally 300 or 375 mm long including a joint. Rebated and socketed flue linings must have ends which fit properly leaving an annular space. Such flue linings should be accurate in shape – Fig 5.11 shows methods of checking these for accuracy.

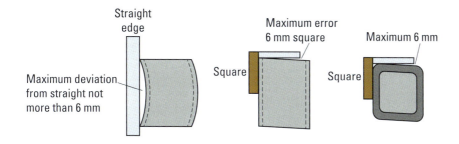

Figure 5.11 Methods of checking accuracy of shapes of flue linings

PRACTICAL TASK

1. CONSTRUCT A CHIMNEY BREAST

OBJECTIVE

To build a chimney to include the standard parts of a constructional hearth and a flue.

Part J of the Building Regulations controls the construction of fireplaces. The main areas are:

* the installation of chimneys and flues for domestic appliances using solid fuel

* BS 1251 specifications for the installation of open-fireplace components

* BS 1181 the specification of liners and terminals (pots)

* Part 1 of the Code of Practice for chimneys and flues.

A fireplace should be built entirely of brick to prevent the spread of fire.

SPECIFICATION

- Two courses to DPC

- Remainder of breast and rear wall to be built as one

- All joints to be solid throughout the wall

- Joint finish to be flush from the trowel to all faces

- Risk assessment to be completed before work commences.

TOOLS AND EQUIPMENT

Walling trowel

Pointing trowel

Spirit level

Lump hammer and bolster chisel

Builder's square

Tape measure

PPE

Ensure you select PPE appropriate to the job and site where you are working. Refer to the PPE section of Chapter 1.

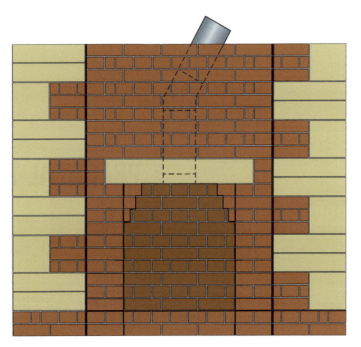

Front elevation

End elevation

Figure 5.12 Plan of chimney breast

Figure 5.14 Plan of course 6

STEP 1 Set out and dry bond the first course on the constructional hearth. Check the attached piers are at 90°. The sides of the fireplace must be a minimum of one brick (200 mm) and extend from the back of the wall by one and a half bricks (338 mm). The width of the opening will depend on the type of appliance that is going to be fitted, but the minimum is two and a half bricks (552 mm).

Figure 5.13 Plan of course 1

STEP 4 The internal area of the chimney breast now needs to be completely closed to form the flue. This can be done in two ways:

Method 1: The brickwork can be corbelled over on each side to reduce the opening to receive the flue liner.

Method 2: Using a purpose made lintel (throat unit) with an opening to accommodate the flue liner.

PRACTICAL TIP

The opening cannot be made bigger after it has been built to accommodate a larger appliance. So if you do not know what type of appliance is going to be fitted, it is better to build the opening slightly bigger.

The constructional hearth is made out of solid concrete and should be at least 125 mm thick. It must extend into the fire opening as well as projecting at least 500 mm in front of the fireplace and 150 mm more than the width of the fireplace opening.

STEP 5 Bed a straight section of flue liner over either the throat unit or the corbelling and continue to build the chimney breast up the top of the liner.

STEP 2 Build the first two courses and lay the damp-proof course.

STEP 6 Using a flue liner that is set at 45°, start the flue coming over. This may need to be done to accommodate another flue coming from a fireplace in the bedroom above.

STEP 3 Continue to build the fireplace until you reach the height of the opening. At this point the opening must be reduced to the size of the flue (225 mm).

STEP 7 Continue to build the chimney breast to the finished height, bedding more flue liners where necessary.

BUILDING DECORATIVE CHIMNEY STACKS

Chimney stacks provide a focal point or feature. It is rare, unless the building is a commercial one, that the chimney is designed to be less visible. In the UK over the centuries there have been an enormous number of different chimney stack designs. These have become key architectural features. The building's period can usually be identified simply by looking at the chimney stack and detailing.

Bonds for decorative chimney stacks

The dividing walls between the flues are called *withes,* and should be of a minimum thickness of 100 mm. To obtain the greatest amount of stability in the chimney stack, the withes should be properly bonded into the outer walling of the stack. Because they are usually 100 mm in thickness the bonding of such stacks necessitates some cutting of the bricks (Figs 5.16 and 5.17).

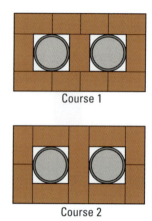

Course 1

Course 2

Figure 5.16 A two-flue stack

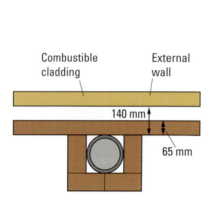

Figure 5.15 The minimum permissible dimensions for a flue in an external wall

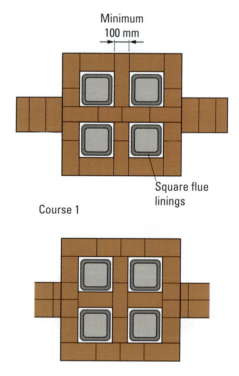

Course 1

Figure 5.17 Plans showing the minimum requirements for the construction of a chimney stack and a method of bonding the brickwork

It is also good practice to use a fairly strong mortar such as Portland cement and sand in a ratio of 1:3 or 1:4 or a masonry cement and sand in a ratio of 1:3.

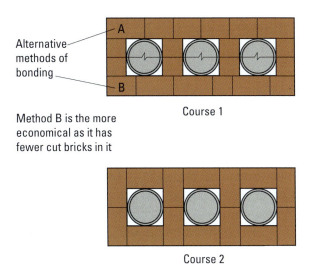

Alternative methods of bonding

A

B

Method B is the more economical as it has fewer cut bricks in it

Course 1

Course 2

Figure 5.18 A three-flue stack

CASE STUDY

Get creative with decorative brickwork

Gary Kirsop, Head of Property Services at South Tyneside Homes, says:

'Decorative brickwork can be so rewarding because you get to take your time to build something that's going to be there for a long time.

During my own apprenticeship, I built a structure for the side of the Jarrow community centre to mark their 50-year anniversary. We had a designer who came and designed it all in clay bricks, which were then taken to Throckley Brickworks where they fired the bricks. The bricks came back and it took us a day and a half to two days to lay them all out and then build it. You get real pleasure from seeing the work completed because it's going to be there for a long time, for people to notice and appreciate.

If you do get the chance to do decorative brickwork, it's about keeping to the tolerances, keeping to the design that you've been asked to make, and making sure you do it correctly. There's a real opportunity here for people who are a bit more creative, to show off their skills. It doesn't always have to be about just building a normal run-of-the-mill wall or block structure. A lot of companies do ask for decorative work to be built. And as a long-term career path, it's certainly something a skilled, creative bricklayer could pursue.'

Single flue and double flue chimney stacks

Single or multiple flues tend to be built into the thickness of a masonry wall. In most cases this means on one of the gable ends. In larger buildings with internal masonry walls, flues need to run up and through the internal walls. In some older properties where there have been flue problems, new flues have been run up externally or internally, again at a gable end.

Flues are normally square or rectangular. The construction of the flue system will depend on the budget set aside for this part of the building work. In many construction jobs, the builders will form an opening for the fire and then continue the flue up within the wall. The size of the openings of the hearth will vary according the use intended.

As fuels have become more efficient over the years, flues have tended to become smaller. In the past, it may have been necessary to try to fit a large number of flues within a gable. However, this has become easier as the flues have become smaller and more efficiency lined.

A withe is the divider between flues in a single chimney stack. The withes can be fairly slender so are very easy to damage. Typically, damage might be caused when the chimney is swept. This is a particular problem at bends and other complex areas of the flue.

Over-sailing and capping

Chimneys need to be made weather proof with mortar flaunching around the chimney pot. The pot goes onto a brick or concrete cap with a drip overhang known as an oversail course. The chimney is usually built using frost-resistant bricks (and with a sulphate-resisting cement). The mortar used for the chimney must be resistant to both rain and frost.

A ventilated cap will prevent water getting into the chimney. These caps can be made from clay, plastic or metal. Caps can be subjected to strong winds, so they must be attached in the right way. If the chimney is unlikely to be used on a regular basis, a ventilated cap will allow sufficient fresh air to move down the entire length of the stack.

The chimney pots will be either circular in cross-section or circular with a square base. The former should be used with circular flue liners and the latter with liners that are square in cross-section.

Chimney pots are usually tapered off at the top to reduce the cross-section of the outlet. This lessens the possibility of downdraught by increasing the upward flow of the flue draught thus preventing the wind blowing down the flue (Fig 5.19).

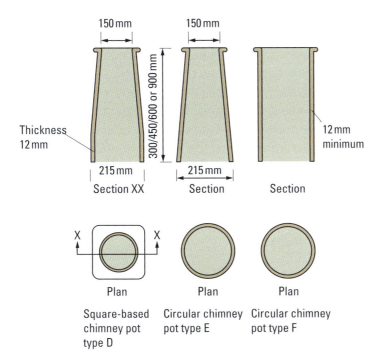

Figure 5.19 Types of chimney pots

The chimney pots must be securely fixed at the top of the stack to prevent dislodgement by the wind, and built well into the brickwork, preferably for at least three courses. They should also project 100 to 175 mm above the flaunching or weathering placed on top of the chimney stack to shed the rainwater. The flaunching may be a mixture of Portland cement and coarse or clean washed sand 1:3 or 1:4. It can be applied in a stiff consistency and trowelled off to an even surface, or finished off with a wooden float (Fig 5.20).

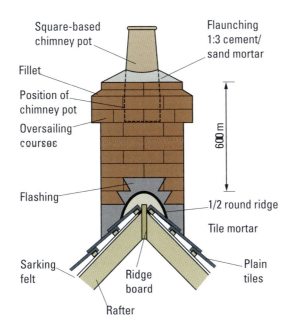

Figure 5.20 A method of building in a chimney pot

Damp-proof barriers

Chimney stacks are very vulnerable to damp because of their exposed position and the junction between the brickwork and the roof. In the case of a flat roof, the method of damp prevention is relatively simple. A damp-proof course is laid across the chimney stack and allowed to protrude from the brickwork. When the roof surface is laid, a skirting is dressed up the chimney stack to a minimum height of 150 mm and tucked into a chase in the brickwork. The chimney damp-proof courses may then be dressed over the top of the roof finish (Fig 5.21).

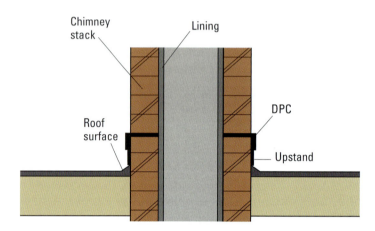

Figure 5.21 Section showing a method of waterproofing a chimney stack where it passes through a flat roof

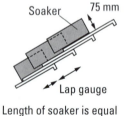

Length of soaker is equal to the lap plus the gauge

Figure 5.22 Detail of a soaker

With a pitched roof the junction between the roof and the chimney stack requires a rather more elaborate method to prevent dampness entering the building.

If the roof covering is of tiles or slates, soakers must be cut from a non-ferrous metal such as copper, lead or zinc. These are equal in length to the gauge of the tile or slate plus the lap and an extra 25 mm for fixing. They are about 175 mm in width, 75 mm of which is turned up to form a right angle (Fig 5.22).

A soaker is placed in between each tile course at an abutment, such as a chimney, and turned up so that it is covered by the flashing. The flashing is fixed into the bed joints of the brickwork. It may be bedded in one bed joint allround the chimney stack or by using the more economical method of bedding it into several joints in line with the slope of the roof. This is called stepped flashing.

The front of the stack is protected from damp penetration by a sheet of metal dressed over the tiles, which is called a front apron. The back of the stack has a chimney back gutter. A cover flashing is fitted to seal the joint between the back gutter and the stack (Fig 5.23).

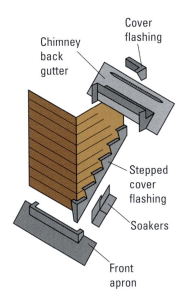

Figure 5.23 An exploded view of a chimney showing a method of waterproofing a chimney stack

As a final protection, a damp-proof course should be placed across the stack (Fig 5.24).

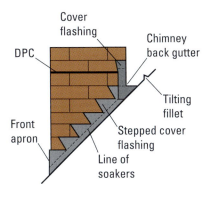

Figure 5.24 Stepped cover flashing

When the flashings are built into the joints there are several points which should be observed so that a perfect barrier is provided against the penetration of rainwater, and also that the damp-proofing material remains securely in place after the completion of the work:

* The joints should be raked out to a depth of about 25 mm.

* They should be raked out cleanly, and not with a sloping section (Fig 5.25).

* When the flashing is in position and plugged in the joints, the bed joints should be carefully brushed out so as to remove any loose particles, and then dampened. The mortar is then trowelled firmly into the joints and pointed on completion.

REED TIP

Customer service experience is great to put on your CV or application form. It demonstrates that you can listen to customers, be polite, and show you understand customers' needs.

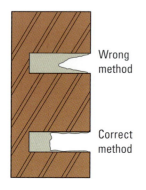

Wrong
method

Correct
method

Figure 5.25 Method of raking out bed
joints for flashing

Jointing

Jointing will depend on the desired look of the brickwork. The timing of jointing will depend on a number of factors:

* Moisture – if the bricks are wet then you will have to wait to joint until they have dried out.

* Temperature – on hot days joints will dry out quickly, as the bricks will absorb heat. On colder and wetter days joints may take time to dry out, as will the bricks.

* Time of day and progress of work – brick courses that you have already laid earlier in the day will be drier than those that you have worked on recently.

* Type of brick – the softer the brick, the more moisture it is likely to soak up, which in turn means that the joints will take less time to dry out.

There are at least six different types of joint finish that you can achieve when pointing. These can be seen in the following table.

Joint finish	Description
Weather struck	The joint slopes downwards to encourage rainwater away from the joint and down the face of the brick.
Ironed or tooled	One of the quickest jointing finishes – a shallow semi-circular indent is made in the joint.
Recessed	During pointing the mortar is compressed into the joint to a depth of around 4 mm.
Flush	The mortar completely fills the joint up to the face of the brick.
Weather struck with cut pointing	The joint angles downwards and protrudes out beyond the face of the brickwork.
Reverse struck	This is the complete opposite of the weather struck joint. The joint angles inwards from the top.

Table 5.1 Joint finishes when pointing

It is also common to use different colour mortar. This can be achieved by adding tints or using different coloured sand. Common mortar colours include the following:

* natural or grey – this is the most commonly used

* white or off-white – this highlights the colour of the bricks

* cream or buff – this tones in with the brick colours

* coloured mortar to match the brick colour – this gives the impression that the whole stack is a solid piece of construction as it is hard to distinguish the individual bricks.

The mortar joint is an important feature of the stack as it shows the banding of the bricks to the best effect. Rolled or ironed joints complement nearly every style and are very popular as they look clean and contemporary. Flush joints will provide a flatter look and will lighten the overall look of the stack by minimising shadows.

Carrying out checks

This should be an on-going process as the job develops. You should ensure that:

* the set out area is correctly located and the footing is checked for conformity to the dimensions and location as per the job drawings and specifications

* the fireplace base is set out to the correct measurements and location to the job drawings

* the base is constructed using the correct mortar mix (which should conform to the specifications)

* the bricks and blocks are laid to the base, conform to the specifications and standards and that they are both laid in line and level

* the damp-proof course has been installed

* bricks, blocks and stone are laid to form hearth to designed shape, pattern, job drawings and specifications.

Setting out brickwork

You should do another on-going series of checks, which should include:

* making sure the face brickwork is laid to form the shape of openings that conform to the designed dimensions and finish of the drawings and specifications

* checking the lintel is installed to specifications

* making sure the brickwork is plumb and level

* ensuring that the bricks that form the outer skin and chimney shaft are constructed in accordance with specifications and standards

* checking the finish of the joints

* checking that the brickwork is raked or ruled to the designed depth in accordance with the job specifications.

CONSTRUCT FIREPLACES AND FLUES

A chimney stack is the terminal for all the flues above the roof level and can contain one or several flues. It can be built very plain or decorative as seen in older buildings. Stacks are open to severe weather conditions so you have to be very careful with the construction to avoid costly maintenance.

The following points should be followed when building any chimney stack:

* Use a suitable brick that can stand exposure to the weather.

* Build the stack to a high standard of finish.

* Make sure that the chimney pot is bedded into position and will not move by high winds.

* The stack should be high enough to clear the roof to discharge the smoke coming from it.

* Flue liners must be carried through the full height of the stack.

* The joint between the roof and the stack should be watertight.

* A lead apron DPC should be inserted 150 mm above the roof line at the front and back of the truss line to prevent water passing down the stack.

PRACTICAL TASK

2. CONSTRUCT A SINGLE-FLUE CHIMNEY STACK

OBJECTIVE

To build the single-flue chimney stack shown in the plan.

SPECIFICATIONS

* Select correct type and quantity of materials

* Build the stack using commons below the roof truss and facing bricks above the line of the truss

* Establish the correct position of the damp-proof course

* Rake out joints to receive lead flashing on all four sides

* Facing joints are to be weather struck

* Risk assessment must be completed before work commences.

TOOLS AND EQUIPMENT

Walling trowel

Pointing trowel

Spirit level

Lump hammer and bolster chisel

Builder's square

Tape measure

Joint rake.

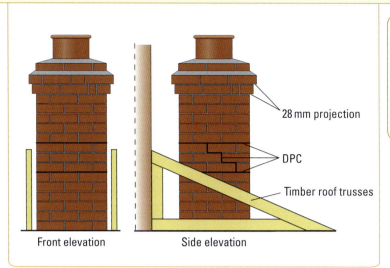

Figure 5.26 Plan of single-flue chimney stack

PPE

Ensure you select PPE appropriate to the job and site where you are working. Refer to the PPE section of Chapter 1.

STEP 1 Set out the work area allocated for this project.

STEP 2 Set out the chimney stack between the two roof trusses and lay the first course. Once the course is correct you can carefully fill in the cross joints.

Plan of course 1

Figure 5.27 Plan of course 1

PRACTICAL TIP

When you are laying each course, this is one of the few occasions when it is best not to use a full perp joint.

If you put full cross joints on, when you tap one side of the pier to get it plumb you will disturb the other brick next to it as there is no give between the two bricks. If the cross joint is not full then the mortar between the bricks will squeeze into the gap and the bricks will not move.

STEP 3 Continue to lay further courses of common bricks until you reach the lower point of the roof truss. At this stage you will have to introduce facing bricks at the front of the stack.

PRACTICAL TIP

Don't forget to put flue liners inside the stack as you are building it.

Flue liners are made from concrete or clay, both of which are non-combustible and resistant to acid. They are either round or square.

They are made with a rebate or socket at one end and a spigot at the other. These have to be installed the correct way to form a perfect seal. The socket or rebate end must be at the top. This stops any moisture in the flue from escaping and getting into the brickwork.

They should be bedded in place by using the same mortar that you are using for the stack or a manufacturer's sealant.

The area between the liner and the stack should be filled with material that will allow any expansion without causing any damage to the liner or stack. This is usually a weak sand/lime mortar or vermiculite concrete.

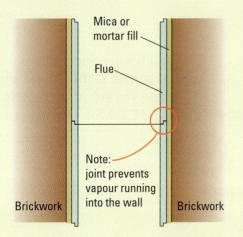

Figure 5.28 Flue liner

STEP 4 Continue to build the stack until it is 150 mm above the bottom edge of the roof truss. At this point you need to insert a lead tray to prevent water penetration down the stack.

The tray is made out of sheet lead and should be at least 50 mm wider on each side of the stack. It must be turned up 25 mm in the inside of the stack to prevent water running down the inside of the stack.

STEP 5 Continue to build the stack until it is 150 mm above the highest point of the truss. Again insert another lead tray as before.

STEP 6 Using the joint raker, rake out the joints to receive the lead flashing around the stack where appropriate.

There are Building Regulation requirements for the termination of the flue and chimney stack at roof level. The main ones are as follows:

* The height of the chimney, at the termination of the flue, and excluding the chimney pot itself must be 1 m above the highest point of contact between the stack and the roof.

* If there is an opening window or skylight in a roof which is less than 2.3 m away (horizontally) from a chimney, then the top of the chimney (again excluding pot) must not be less than 1 m above the top of the window.

* If a roof has a chimney within 600 mm of the ridge, then the chimney must not be less than 600 mm above the ridge (on roofs over 10° pitch angle).

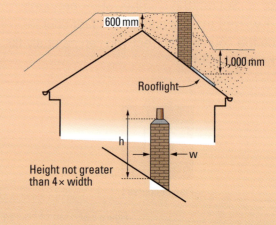

Figure 5.29 Building Regulations showing minimum height

STEP 7 Continue to build the stack until the required height to the oversail course.

STEP 8 Lay the first course of the oversail.

Oversail courses are used to protect the top of the wall from the weather. Along with a mortar fillet they allow rainwater to run off the top of the wall and away from the face of the main wall.

Make sure that you get the projection the correct size, and that none of the bricks are tipping. Make sure that you project the bricks by at least 25mm.

STEP 9 Lay the second oversail course as before.

STEP 10 Bed the chimney pot in position. Make sure that it is secure and will not blow off in strong winds.

Figure 5.30 The chimney pot

STEP 11 Set the top course of bricks back to the original size.

STEP 12 Complete the top of the stack by putting a mortar fillet around the two oversail courses.

PRACTICAL TIP

The flaunching allows any rain water to run off the top of the pier and prevents it from getting into the joint. If water was allowed to seep into the joint, in winter it would freeze and the brick would become loose.

PRACTICAL TASK

3. CONSTRUCT A DOUBLE-FLUE CHIMNEY STACK

OBJECTIVE

To build a double-flue chimney stack.

See the previous practical task for specifications, tools and equipment and PPE.

STEP 1 Set out the chimney stack between the two roof trusses and lay the first course. Once the course is correct you can carefully fill in the perp joints.

STEP 2 Follow steps 2 to 12 of the previous practical task. Construct a single-flue chimney stack.

Plan of course 1

Figure 5.31 Plan of course 1

TEST YOURSELF

1. Which of the following would **not** be found in a Bill of Quantities?

   a. Architect's working drawings

   b. Specification for the job

   c. Schedule for the job

   d. your timesheet

2. Which of the following statements about chimneys, flues and hearths is **untrue**?

   a. Materials used should be of a non-combustible nature

   b. They need to be thin enough not to retain heat

   c. They need to prevent smoke escaping into the building

   d. They should not be affected by condensation

3. If an oil-fired appliance was being fitted into a fireplace, which organisation might you approach for a list of registered members?

   a. HETAS

   b. CORGI

   c. OFTEC

   d. NACS

4. If the chimney is square or rectangular, what is the maximum size of the fireplace opening compared to the area of the flue?

   a. 8 times

   b. 4 times

   c. 10 times

   d. 12 times

5. What is the minimum thickness recommended for a hearth that has an appliance fitted into it?

   a. 50 mm

   b. 100 mm

   c. 125 mm

   d. 150 mm

6. When sulphur dioxide or carbon dioxide mixes with water what is produced?

   a. An acid

   b. An alkali

   c. A paste

   d. A solid

7. What is the minimum height that a chimney top should be from the ridge line of a building?

   a. 250 mm

   b. 600 mm

   c. 900 mm

   d. 1,200 mm

8. What are the dividing walls between flues called?

   a. Stays

   b. Blocks

   c. Withes

   d. Cavities

9. What materials are flue liners made from?

   a. Concrete or uPVC

   b. Clay or uPVC

   c. Only concrete

   d. Concrete or clay

10. Which type of joint finish can be described as being created by a shallow, semi-circular indent?

   a. Ironed or tooled

   b. Weather struck

   c. Recessed

   d. Flush

Chapter 6

Unit CSA–L3Occ133
CONSTRUCT COMPLEX MASONRY STRUCTURES

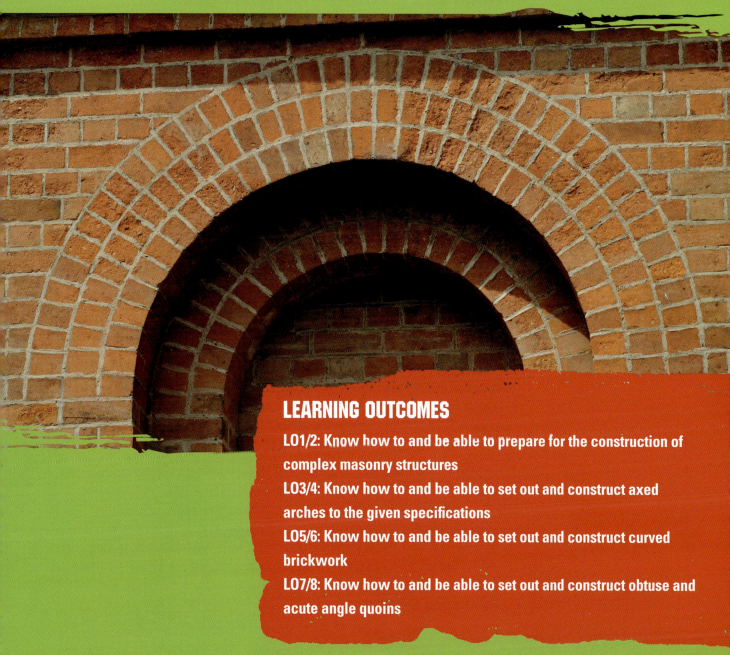

LEARNING OUTCOMES

LO1/2: Know how to and be able to prepare for the construction of complex masonry structures

LO3/4: Know how to and be able to set out and construct axed arches to the given specifications

LO5/6: Know how to and be able to set out and construct curved brickwork

LO7/8: Know how to and be able to set out and construct obtuse and acute angle quoins

INTRODUCTION

The aims of this chapter are to:

* help you to select materials, components, tools and equipment
* help you to construct complex masonry structures.

PREPARING FOR THE CONSTRUCTION OF COMPLEX MASONRY STRUCTURES

Preparing for the construction of complex masonry structures requires much the same considerations as preparation for any other type of building. This includes consideration of:

* potential hazards
* types of drawings and specifications and how to interpret and extract accurate information from them
* risk assessments
* PPE
* protecting materials, completed work and the environment.

Health and safety and hazards

In Chapter 1 we examined the importance of identifying potential hazards and the ways in which risk assessments and method statements can help to avoid possible health and safety hazards. It is also important to make sure that manufacturers' instructions are followed. All of these precautions help to ensure that you follow health and safety legislation and keep yourself safe.

In Chapter 1 there is also information about Work at Height Regulations and how to ensure that any equipment and safety measures you use will protect you. You should always try to avoid working at height if possible. Always use the proper equipment and put measures in place to prevent falls if this cannot be avoided. There will always be a risk, even if the proper equipment and safety measures are taken. As there is always a risk, measures need to be put in place to minimise the distance and consequences of a fall.

Checking drawings and specifications

Each type of drawing has a specific purpose and together they should provide you with the full picture of exactly what is required.

The purpose of drawings is to assist construction. They are organised in a logical sequence, which should follow the flow of the actual construction work. Scales should be clear and show the details of the structure.

Drawings and their purpose are covered in detail in Chapter 2.

It is important to ensure that all documents used are accurate and that each matches the other.

All drawings must follow the requirements of BS 1192:2007. This means that the drawings will have a common format and symbols. Building drawings use what is known as first angle orthographic projection. Drawings in this form will have a special symbol.

Interpreting measurements

In Chapter 2 we detailed the types of symbols and hatchings used on drawings and specifications. These are industry-standard and they help you to interpret the materials required for the job in hand.

Reporting inaccuracies

If, at the outset of a job, it becomes obvious that there have been some inaccuracies in the documents prepared for the construction work, it is vital that this is reported immediately. It may be necessary to report any inaccuracies to the site manager or, if appropriate, to an architect or designer.

A bricklayer is unlikely to be responsible for either surveying or setting out the project. However, a supervisor does have the responsibility to check to ensure that the work follows what has been outlined in the drawings. It means knowing if something has gone wrong or could go wrong.

Once the setting out has been carried out it is good practice to double check. Additional checks should be carried out before important new stages of construction get underway. In any case of a discrepancy the drawing needs to be checked to ensure it is correct. This means almost certainly talking to the architect or designer to clear up any misunderstandings and avoid potential problems.

Potential hazards

Typical construction potential hazards can include:

* falls from scaffolding, ladders and roofs

* electrocution

* injury from faulty machinery

* power tool accidents

* being hit by construction debris

* falling through holes in flooring

* fires and explosions

* burns, including those from chemicals

* slips, trips and falls.

PPE

PPE should include the following:

* Safety goggles or safety glasses – these are always necessary if there is a risk of eye injury.

* Gloves – these are especially useful if you are working in cold conditions or regularly using cement, solvents or other potentially harmful chemicals.

* Safety footwear – in case of sharp objects, uneven ground and poor weather conditions.

* Hard hat – to protect your head from objects falling from above and from bashing it on structures.

* High visibility clothing (hi-vis) – these may be waistcoats, jackets, overalls, or clothing with reflective pads, to ensure you can be easily seen on site.

Resources

The specification will tell you what types of materials are to be used. They will also tell you the joint finish required.

Using the plans you can work out the materials you will need for the job. You will also need a range of brickworking tools and PPE. Before you get underway you need to check that your materials and your equipment are ready for use. You should always ask yourself the following questions:

* Do I have the right materials and are there enough of them?

* Are the materials in good condition or are they damaged?

* Do I have all the tools that I will need for the work and are they in a good state of repair?

* Do I have the necessary PPE and has it been maintained properly?

Calculations and formulae for quantities

In order to work out the amount of materials required for a particular walling job, you need to know the area of the wall. This is the surface that it covers. This is simply achieved by multiplying the length by the width. This gives the area of the walling.

For some jobs it can be more complicated, as there may be separate walls of different shapes. In this case you need to work out the area for each shape and then add them all together.

Working drawings will usually show lengths and widths in millimetres, so these need to be converted into square metres. It is often a good idea to convert from millimetres to metres before you start making any calculations. See Chapter 2 for more on drawings and working out areas.

The next stage is to work out how many bricks you will need for each square metre of walling. As a general rule of thumb, there are 60 bricks per square metre of half-brick walling. Once you know the total area of the wall that you will be constructing you multiply the number of square metres by 60.

Protecting the work and the surrounding area from damage

Newly built brickwork can be vulnerable to cold temperatures and rain. The brickwork should be encouraged to dry out, which may mean putting a covering material over the face of a wall. Hessian is also used as an insulating layer.

Waste is unavoidable but strict environmental legislation determines how you handle and dispose of waste. The waste that is produced will cause some environmental damage and this must be minimised by:

* reusing broken bricks and blocks as hardcore

* sweeping fine debris into heaps and sprinkling water on it to minimise the dust

* not returning unused mortar to be remixed as this is bad practice

* not burning waste

REED
TIP

Interviews are about Preparation, Presentation, and Personality.

Do some research about the companies or organisations you're applying to work for.

Turn up smartly dressed.

Talk about things that interest you and why you're passionate about becoming a bricklayer.

* keeping all material bins shut or locked when not being used

* bagging up any waste, or at least covering it up.

SETTING OUT AND CONSTRUCTING AXED ARCHES

A more ornamental way of bridging openings is by the use of arches, which are made up of bricks bonded together around a curve or series of curves. Arches need no additional reinforcement, unlike brick lintels, because the shape of the arch distributes the load sideways and around the arch. The higher the load placed on the arch, the tighter it will become.

Arches are generally classified into three main groups according to their use or method of cutting and preparation:

* Rough arches – where the joints, and not the bricks, are wedge-shaped. Such arches are generally used on work which does not require a high standard of finish, or work which is to be plastered over. Little or no cutting is needed for this type of arch.

* Axed arches – are used on fine work, and the wedge-shaped bricks called voussoirs are all cut to the same shape and size and have a pleasing appearance when finished. They may be cut from the same bricks as the general face work, or sometimes bricks which have contrasting colour are used to the effect of the arch standing out from the general walling.

* Gauged arches – which are very ornamental and expensive and are generally only used on higher quality buildings as the bricks require a lot of preparatory work before they can be built in the arch. They are prepared and bedded with a very fine white joint. Building this type of arch requires a sound knowledge of geometry.

Axed arches terminology

There are a number of terms related to arches that you need to be familiar with. The following table identifies the most commonly used terminology.

Term	Explanation
Voussoirs	The individual wedge-shaped bricks in an arch.
Span	The distance between the jambs or reveals of the opening over which the arch is bridged.
Soffit	The surface of the underside of the arch.
Springing points	The lowest points of the arch from where the curve begins.
Springing line	A horizontal line drawn through the springing points.
Rise	The vertical distance between the springing line and the highest point of the soffit.
Key brick	The highest or central brick. It is usually the last brick to be laid in the arch.
Crown	The highest part of the arch, where the key brick is laid.
Intrados	The underside of an arch when viewed in elevation.
Extrados	The upper side of the arch when viewed in elevation.
Haunch	The lower third of the arch, from the springing line to midway to the crown.
Bed joints	The joints between the voussoirs.
Face joints	The cross joints between the voussoirs.
Skewback	The sloping abutment from which an arch springs (see Fig 6.2).
Template	This piece of material is marked out from the drawing of the arch and cut to the shape of the voussoirs. The bricks are cut to match the shape of the template.
Striking point	This is the centre point from which the curve of the arch is drawn. The voussoirs should radiate to the striking point.
Turning piece	A solid piece of wood in the shape of the arch. It is used as a support.
Arch centre	This is another method of support while the arch is being constructed. (We look in more detail at providing temporary support a little later in this section.)

Table 6.1 Terms related to arches

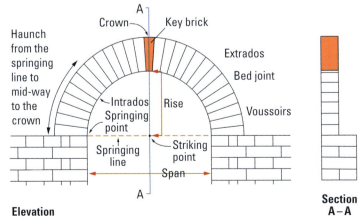

Figure 6.1 Semi-circular arch

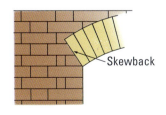

Figure 6.2 Skewback

There are several different types of common arch. These are explained in the following table.

Type of arch	Description
Semi-circular	If the brickwork is not going to be seen, the arch is made using standard bricks and the mortar joints are v-shaped joints. If the brickwork will be visible then a central key brick is used with an equal number of specially shaped voussoirs either side.
Segmental	This is called a segmental arch because only part of a full circle is actually constructed. All of the voussoirs are the same. Usually these arches will have a low rise and a large span. This means that standard bricks can be used, which are held in place with v-shaped joints. To make the arch more visually appealing, particularly if it is a smaller span arch, tapered voussoirs are used. There is always an odd number of voussoirs and the central one is the key brick.
Soldier	This looks like an arch with a straight top. It is non-structural and built by fixing bricks on their ends, which in turn are supported by a steel lintel. It looks like an arch but the bricks do not actually perform any supporting function.
Flat or skewback	The voussoirs in this type of arch are either bought to order or are cut on site from a template. Each of the voussoirs has a unique shape. The arch is flat at the top and the ends slope (skewback). Effectively this creates a wedge that supports the structure.

Table 6.2 Different types of common arch

Setting out axed arches and providing templates

The following is an outline of the general procedure for preparing the voussoirs for an axed arch. This procedure may be slightly varied to suit the different types of arches but the basic principles remain the same.

1. An outline of the arch should be set out full size on a setting-out board.

2. Mark out the voussoirs on the extrados of the arch. The voussoirs are set out by marking out the key brick first then dividing the extrados into a number of equal divisions. The marked divisions should not be greater than the width of the bricks being used plus the thickness of one bed joint. If the arch is to be bonded on face there must be an even number of voussoirs on each side of the key brick, as seen in Fig 6.3. This will ensure that the springing brick corresponds to the key brick.

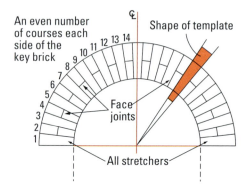

Figure 6.3 There must be equal numbers of voussoirs either side of the key brick

3. When the correct size of the voussoirs has been determined, the shape of the arch can be completed. This is done by drawing the joint lines between the extrados and the intrados so that they radiate to the striking point of the curve. This ensures that the bricks are normal to the curve (square to the tangent at the point on the curve it passes through), as can be seen in Fig 6.4.

You can now cut out a template. This can be made from plywood or hardboard and is cut to the shape of the voussoirs on the full size drawing. One method of marking the template is to extend any two lines of the arch voussoirs, laying the plywood or hardboard between and transferring the lines to the upper surface by means of a straight edge. The template should be considerably longer than the depth of the arch and should extend at the narrow end of the template, as can be seen in Fig 6.5.

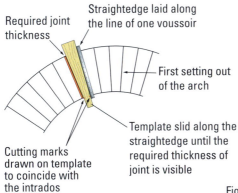

Figure 6.5 Setting out an arch

Figure 6.4 The voussoirs need to be normal to the arch curve

Check the accuracy of the template shape against other voussoirs in the full size drawing.

After the shape is checked the joint can be determined by laying a straight edge along one edge of the voussoirs on the drawing and placing the template against it so that it fully covers the voussoir. Then move the template along the straight edge until the required thickness of bed joint is visible on the drawing. Mark on each side of the template the cutting mark for the bricks. It is useful to nail a strip to the template at the cutting mark, as can be seen in Fig 6.6.

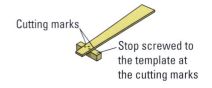

Figure 6.6 A template for arch voussoirs

Mark the bricks by placing the template on the face of the brick, as shown in Fig 6.7. Mark the soffit with the aid of a square, as shown in Fig 6.8.

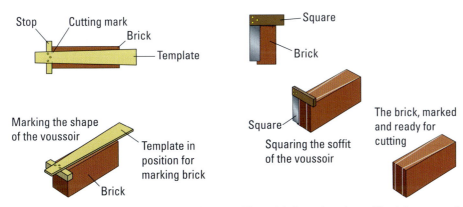

Figure 6.7 Marking the voussoirs   Figure 6.8 Squaring the soffit of the voussoir

Providing temporary support

When construction is taking place it is necessary to provide some form of support to the brickwork over the opening. As we have already seen, no permanent reinforcement is required so any support will be carefully removed once the arch has been built. Two common methods of temporary support are the turning piece and the arch centre.

Turning piece

A turning piece is formed from solid pieces of wood into the actual shape of the arch. Turning pieces are generally used for segmental arches which have only a small span and rise. It is more economical to build up a centre from a small piece of timber when arches of dimensions are greater than 1 m span and 75 mm rise are to be formed.

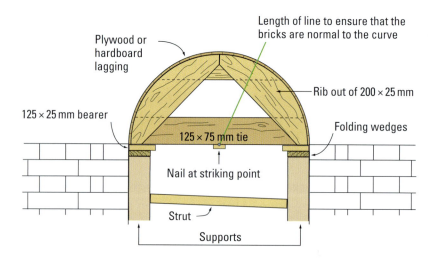

150 mm

100 mm

Figure 6.9 A turning piece for a segmental arch made from two pieces of 100 by 75 mm wood nailed together

Arch centre

An arch centre is made up of a number of small section timber pieces formed into the shape required.

Plywood or hardboard lagging

Length of line to ensure that the bricks are normal to the curve

Rib out of 200 × 25 mm

125 × 25 mm bearer

125 × 75 mm tie

Folding wedges

Nail at striking point

Strut

Supports

Figure 6.10 The arch centre in place

Arch centres can be conveniently used over quite large spans. The timber members of a centre include the following:

* Ribs – these are the shaped members which are formed to suit the required arch shape.

* Ties – these are placed across the lower part of the centre to prevent it from spreading out when it is carrying the weight of the arch.

* Laggings – these are small pieces of timber that are fixed across the ribs to carry the voussoirs. Centres may be open or close-lagged, although open lagging is an inferior method because of the difficulty in marking the position of the voussoirs on the centre before building the arch. An alternative method is to use resin-bonded plywood nailed on to the ribs instead of the small timber laggings. This provides a smooth surface on which to work and is the most efficient method (see Fig 6.11).

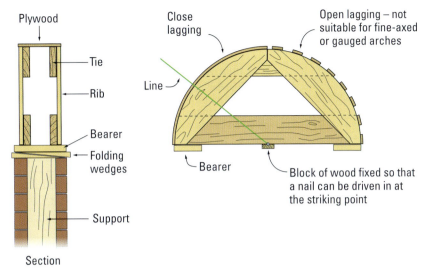

Figure 6.11 A typical arch centre

- Bearers – these are the timbers which are fixed underneath the ties to prevent the ribs from spreading apart. They also carry the weight of the centre and arch.

- Struts – these are used in larger centres to support the ribs off the centre and go between the props (see Fig 6.10).

- Props – these are the main supports to keep the centre at its correct height.

- Folding wedges – these are placed between the props and the bearers to provide any slight adjustment that may be necessary to the height of the centre before beginning to set the arch in position.

Construction methods and procedures for arches

Once the template has been made, and the temporary supports have been placed, then the construction of the arch can begin. The procedure for construction is given below:

- Cut the bricks using a hammer and bolster.

- Secure the centre of each voussoir carefully in position. You can make any adjustments required using the folding wedges.

- Accurately mark the position of the voussoirs on the intrados and the width of the joints.

- Drive a nail at the striking point(s) and attach a length of line so that the radiation of the bricks to the striking point can be checked for accuracy and that they are normal to the curve of the arch.

- Check the arch for straightness along its face. This can be done by building up the brickwork each side of the arch and stretching a line in between (see Fig 6.12). Alternatively a line can be stretched between two temporary one-brick piers that have been erected on each side of the opening. These temporary piers are known as dead men (see Fig 6.13).

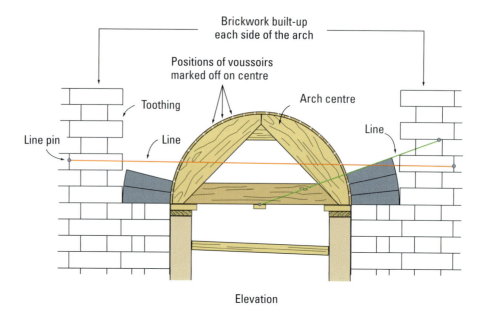

Figure 6.12 Building up the walling each side of the arch

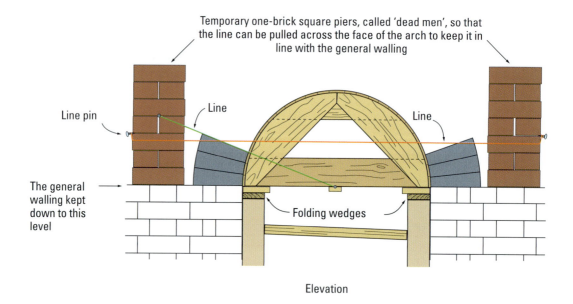

Figure 6.13 Dead men can be used to check for straightness

* Build up the arch evenly on each side, meeting it at the middle or key brick.

* Check each voussoir for its correct position on the centre and correct alignment by means of the line from the striking point.

Constructing a semi-circular and a segmental arch

The method of constructing these two different arches is very similar.

* The opening needs to have a tolerance of around 5 mm.

* The centre of the arch needs to be identified.

* The abutments should be built up to the springing line. The centre is supported by timber props and wedges. At all times they should be checked for being level.

* The corners should be built up to 4 or 5 courses.

* For segmental arches, the skewback angle is marked on the supported bricks. These are then cut and bedded in as each course of bricks is laid. This continues until you reach the top of the skewback angle.

* A line is pulled through the corners and the face of the voussoirs is laid to the line.

* Each voussoir is levelled as it is laid. It is important to check that they are laid square.

* Voussoirs on each side of the arch are laid carefully, with full joints; otherwise the arch will be weak.

* Brickwork is then laid in courses around the arch, making any necessary cuts to cope with the shape.

* As the brickwork around the arch is built up the key brick is then laid, with well-filled joints.

* When the arch is complete the joints should be allowed to contract as they set. Do this by gently easing the wedges to allow the arch to settle. Any supports are removed after the joints have set and the soffit of the arch is jointed.

Safely removing the support

Providing the arch has been allowed to set, the temporary supports can now be removed. The folding wedges can then be removed completely. Then the centre can be taken out.

1. CONSTRUCT A SEMI-CIRCULAR AXED ARCH

OBJECTIVE

To construct a semi-circular axed arch, using an arch former and a voussoir template.

In an axed arch, the bricks are shaped so that the bricks are consistent and regular. This creates a neater finish than a rough arch, where the joints dictate the way that the arch is formed. To achieve this, the arch has to be set out accurately and each brick marked and dressed as identical units.

This model is for a semi-circular arch, with the main body of the wall constructed in two stretcher bond walls built adjacent to each other so that you can concentrate on the arch itself.

Figure 6.14 Elevation of a semi-circular arch

TOOLS AND EQUIPMENT

Walling trowel

Lump hammer

Bolster chisel

Spirit level

Block/pins and line

Straight edge

Jointing iron

Steel tape measure

Compass dividers

PPE

Ensure you select PPE appropriate to the job and site where you are working. Refer to the PPE section of Chapter 1.

STEP 1 Mark a straight line on the floor and dry bond the wall 890 mm long at each side of the arch centre.

STEP 2 Dry bond the bricks either side in order to build up the corners and build the piers up to the springing point of the arch.

PRACTICAL TIP

Make sure that the bond will work throughout the length so that the correct bond is maintained when the brickwork runs above the arch.

STEP 3 To set out the arch, draw the span on a piece of board then place the piece of board on the face of the arch to mark the voussoirs on. Then bisect the span in order to find the centre point.

STEP 4 Mark the intrados and extrados on the board with a trammel, so that the voussoirs can be marked.

STEP 5 Take a pair of dividers so that the width of the voussoirs can be established, and mark on the board.

PRACTICAL TIP

Start with the dividers set at the width of a brick (65 mm) and adjust by reducing until the spacing around the extrados divides equally.

STEP 6 To make the voussoir template:

Radiate the lines to the centre point and extend them to above the extrados.

Transfer the lines onto the hardboard template and cut out.

Place the template on one voussoir and position straight edge along one edge.

Slide the template along the straight edge until the right sized joint is achieved.

Mark the position and fix a small batten to fix the size of the cuts to be made.

STEP 7 Cut the bricks and lay flat on the board, ready to be laid to the arch.

Figure 6.15 Cutting the voussoir

PRACTICAL TIP

Scribe the cuts with a sharp steel point rather than a pencil. Use a set square to transfer the lines on the face to the edge of the brick.

STEP 8 Place the arch centre. Ensure that it is plumb and does not project beyond the face of any further brickwork.

STEP 9 Build the corners up as far as possible so that the alignment of the arch can be maintained as its constructed.

STEP 10 Start building the voussoirs over the former with a constant joint and maintain alignment with the string line. The position of the voussoirs can be checked with a line attached to the springing point. Work from either end to the centre.

Figure 6.16 Building the voussoir

STEP 11 Complete the arch and run the full width courses over the arch.

STEP 12 Finish the model with a half round joint and brush it.

2. CONSTRUCT A SEGMENTAL AXED ARCH

OBJECTIVE

To construct a segmental axed arch, using an arch former and a voussoir template.

This model is for a segmental arch with the main body of the wall constructed in two stretcher bond walls built adjacent to each other.

Figure 6.17 Elevation of a segmental axed arch

TOOLS AND EQUIPMENT

Walling trowel	Straight edge
Lump hammer	Jointing iron
Bolster chisel	Steel tape measure
Spirit level	Compass dividers
Block/pins and line	Adjustable bevel

PPE

Ensure you select PPE appropriate to the job and site where you are working. Refer to the PPE section of Chapter 1.

STEP 1 Follow Steps 1 and 2 from Construct a semi-circular arch, above.

STEP 2 After you have marked the voussoirs, establish the span and bisect it. Then bisect the line from the rise to the springing point.

STEP 3 Establish the intrados, extrados and skewback and mark on the board.

STEP 4 Take a pair of dividers so that the width of the voussoirs can be established, and mark on the board.

Figure 6.18 Marking the width of the voussoirs

PRACTICAL TIP

Start with the dividers set at the width of a brick (65 mm) and adjust by reducing until the spacing around the extrados divides equally.

STEP 5 Follow Steps 6 to 8 from Construct a semi-circular arch, above.

Figure 6.19 Using an adjustable bevel to get the angle of the scewback

Figure 6.20 Measuring out the voussoir template

Figure 6.21 Using the voussoir to mark out the angle

Figure 6.22 Trimming the voussoir

STEP 6 Construct the skewback ready to take the voussoirs over the former.

STEP 7 Build the corners up as far as possible so that the alignment of the arch can be maintained as it is constructed.

STEP 8 Start building the voussoirs over the centre with a constant joint and maintain alignment with the string line. The position of the voussoirs can be checked with a line attached to the springing point. Work from either end to the centre.

Figure 6.23 Close up of the voussoir

PRACTICAL TIP

Don't be tempted to build the arch above the height of the corners. Build a section of arch and infill with brickwork and then repeat the process until over the arch.

STEP 9 Complete the arch and run the full width courses over the arch.

STEP 10 Finish the model with a half round joint and brush to finish.

Figure 6.24 The completed arch

SETTING OUT AND CONSTRUCTING CURVED BRICKWORK

Many modern buildings incorporate curved walling or brickwork features, including bay windows and garden walling. Curved brickwork, like an arch, will have a striking point, or centre, from which all of the brickwork will radiate. Vertical plumbing and horizontally levelling are very important when building curved brickwork. It is not possible to use lines and pins for curved walling.

Setting out curved brickwork on plan

This involves the use of one of three methods:

* Wooden templates – commonly used on bay windows, they give the line of the main wall and the curve of the bay. They are used when the curve of the bay has a small radius.

* Templates and plumbing points – when the radius of the curve is larger, the template is used to lay only the first course of brickwork. Plumbing points are then marked around the curve. As each course of brickwork is started a brick is laid at each of the plumbing points. These bricks are bedded in bond and levelled horizontally from the main wall and plumbed vertically from the first course of bricks.

* Trammels – these are pieces of steel rod or conduit that are fixed into position and plumbed. A batten is drilled so that it fits easily over the rod and is cut to the length of the radius. The batten is threaded over the rod and the wall can be built to the batten.

Construction methods and procedures

When the radius of the curved work is large enough the bricks can be laid around the curve with 'V' joints between them so that no cutting is necessary on the front 112 mm of the facing course of a one-brick wall. A certain amount of cutting will, however, be required on the inside of the curve because of the reduced radius (see Fig 6.25). The face work may be built using any of the principal bonds.

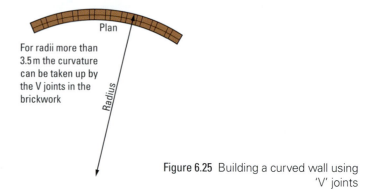

Figure 6.25 Building a curved wall using 'V' joints

If the face work is on the concave face (see Fig 6.26) it may be difficult to do the bonding on the convex side of the wall. If both sides of the wall should be neat work it is essential to be extremely careful and it is normally better to use purpose-made bricks. These may be supplied as headers or stretchers and will allow a good face to be obtained on both sides of a 225 mm (one-brick) wall, for example.

Another method of building curved work, particularly when the radius is small, is to use all heading bond on the face side of the wall. If the radius is very small, bats will have to be introduced (see Fig 6.27).

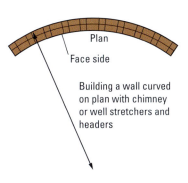

Building a wall curved on plan with chimney or well stretchers and headers

Figure 6.26 Walls curved on plan

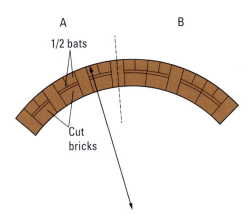

Figure 6.27 Alternative method of using heading bond for a wall curved on plan

Serpentine walling

This type of walling curves in and out along its length (see Fig 6.28).

Its use is normally confined to boundary walling and gives a pleasant non-monotonous effect. It also may be seen on some large housing estates where the roads have been deliberately constructed with curves in them to control the speed of the traffic. Low serpentine boundary walls are then constructed parallel to these roads.

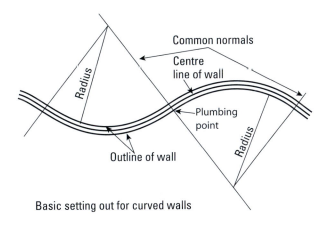

Basic setting out for curved walls

Figure 6.28 Serpentine walling

When building walls which are curved on plan, it is most important to set out the plumbing points at the base and to maintain these points all the way up the wall. The work in between the plumbing points should be checked by the use of a template cut to the shape of the curve, out of plywood or hardboard (see Fig 6.29).

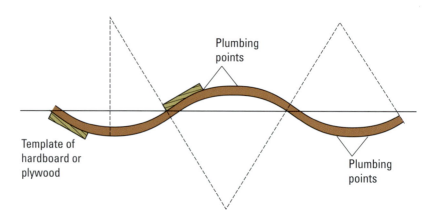

Figure 6.29 Plan showing a method of maintaining the true alignment of a serpentine wall throughout its height

Another method for checking accuracy is to use a trammel, as can be seen in the following diagram.

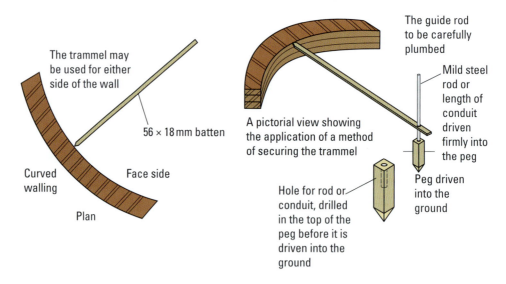

Figure 6.30 The use of a trammel for the building of a concave curve

PRACTICAL TASK

3. BUILD A SERPENTINE WALL

OBJECTIVE

To practise the skills of building a serpentine wall as well as those for the features that are constructed within the pier.

One of the unusual aspects of this model is the skill required to plumb a wall along an irregular length, as a line cannot be utilised. Different techniques for maintaining regularity can be used, but careful checking is essential.

TOOLS AND EQUIPMENT

Walling trowel	Scutch hammer
Lump hammer	Steel tape measure
Bolster chisel	Gun template
Spirit level	Boat level
Jointing iron	Steel or timber trammel

Plan of course 1

X – plumbing points

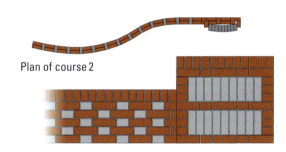
Plan of course 2

Figure 6.31 Plan of a serpentine wall

PPE

Ensure you select PPE appropriate to the job and site where you are working. Refer to the PPE section of Chapter 1.

STEP 1 Mark the position of the model on the floor, starting at the pier end.

STEP 2 The regularity of the serpentine wall can be maintained by fixing a pin at a certain point and using a line or trammel to set the arc. This can then be checked by the use of the template.

STEP 3 Build the pier section, making sure to leave the recess for the feature on the piers to be built in as work progresses.

PRACTICAL TIP

Place some support under what will become the feature, preferably with a piece of timber directly under it so that the arc can be calculated and marked as a reference.

STEP 4 Install the soldiers in the arc, using a boat level to check for plumb and a template to maintain the arc correctly.

STEP 5 Rack out the bricks into the serpentine wall enough to be able to reach the height of the pier.

STEP 6 Build in the second feature by supporting the projecting brickwork and repeating the process that was undertaken with the first.

STEP 7 Finish the pier to height but leave off the main capping on the wall until the remainder of the wall has been built.

STEP 8 Mark six equally spaced plumbing points on the floor, along the face of the serpentine wall.

STEP 9 Lay the first course of the serpentine wall using the spirit level to check the wall for level along the width and length of the wall. Check for alignment with the template.

PRACTICAL TIP

A straight edge can be used in conjunction with the spirit level to check the entire course once it has been laid.

STEP 10 Continue running out each course using the spirit level along with the plumbing marks as reference points and the template to check the accuracy of the radius.

STEP 11 Run the brick-on-edge capping on the pier section. All the bricks can be laid in order, or the full bricks can be laid first and the cut bricks bedded in after these have been completed. Each method is down to individual preference.

PRACTICAL TIP

Running the upper level of copings in first will help prevent the lower section becoming marked.

STEP 12 Run the brick-on-edge capping in using the same method as for the body of the serpentine wall.

STEP 13 Finish the main wall with a half round joint and a flush joint to the brick on edge cappings.

CASE STUDY

LAING O'ROURKE

Time and experience count for a lot

Marcus Chadwick, a bricklayer at Laing O'Rourke talks about building complex structures.

'At college, you probably only get to do one project on complex structures, so you only really start learning on the job. You've got to take notice of the people who are teaching you. Not only your tutors – although they're giving you an insight – but you only really start learning once you're on site, where you've got your old boys who've been doing it for decades. Especially when you consider that they didn't have as much mechanical help like forklifts – it was more hands on, whereas we're very mechanically orientated now. So it's by paying attention to those more experienced people that you learn about things like building arches. It's a dying part of the trade. If I hadn't paid attention to it, I wouldn't have known how to do it and I wouldn't have been able to pass it on.

The basis of your arch is the timber arch-former (centre), and it's about knowing exactly where to place those bricks so that it'll work out when you get to the top and you put your keystone in. That keystone, once that arch-former comes out, that's what's holding that arch together – and gravity and mortar of course. Remember, some of these arches have been up for hundreds of years – so the old masons must have known what they were doing!

I like doing that sort of more complex work, it's satisfying. But don't get me wrong, I love coming to work with a nice long stretch of brickwork to do. I feel really good if I can put down 500 or 600 bricks in a day. And you do end up counting your bricks!

I used to work with this old bricklayer who was 72 – we were taking the mickey out of the young lads because between us we were doing over a thousand bricks a day and the young lads were doing like 400 to 500 between two of them. Pretty impressive when you consider that my mate's 72. But you develop speed with time and experience – it's no good going out on a site and trying to keep up with bricklayers, because you'll just end up taking your work down. And once that's happened, it's a bit embarrassing, so it'll only ever happen the once!'

SETTING OUT AND CONSTRUCTING OBTUSE AND ACUTE ANGLE QUOINS

A quoin is a corner brick. They can provide strength for walling, or can be used as a feature of a corner. Brick quoins may protrude from the facing brickwork on a building to give the appearance of blocks.

Setting out and constructing obtuse and acute angles

Not all sites are square on plan, particularly in town or city areas and, therefore, it is often necessary for walls to be built out of square with one another. Walls that are set at an angle of less than 90°, or a right angle, form an acute angle.

There will be little difficulty in cutting the shapes of the bricks at the angle if a mechanical brick-cutting saw is available. It is unlikely that this work would be cut by hand because of the costs involved in hours.

If an acute angle is formed in an exposed position it is liable to become damaged. In such cases it is usual to cut the angle short and to form two obtuse angles, as can be seen in the following diagram.

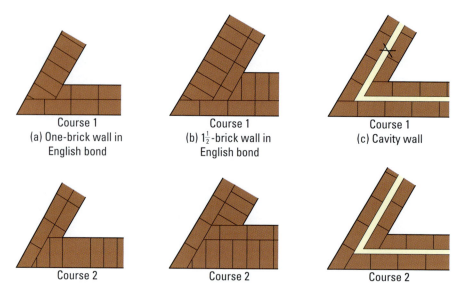

Course 1
(a) One-brick wall in English bond

Course 1
(b) 1½-brick wall in English bond

Course 1
(c) Cavity wall

Course 2

Course 2

Course 2

Figure 6.32 Bonding acute angles in English bond

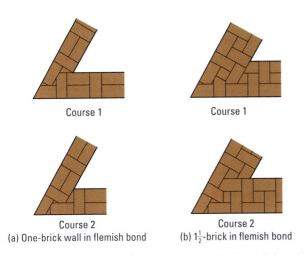

(a) One-brick wall in flemish bond

(b) 1½-brick in flemish bond

Figure 6.33 Bonding acute angles in Flemish bond

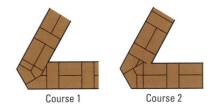

Figure 6.34 An alternative method of bonding an acute angle in Flemish bond

When obtuse angles are being formed, it is possible to use purpose-made bricks. These are generally used for standard angles, such as 135° or 120°. For other angles the bricks would have to be purpose-made if a sufficient number were required. If only a small number were needed they would be cut on a mechanical brick-cutting saw.

It is important to remember that a closer should always be placed next to a squint brick to form the bond in the brickwork, as shown in Fig 6.35.

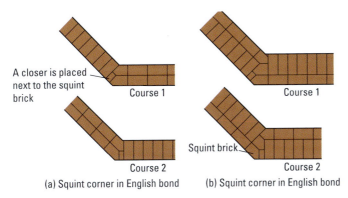

(a) Squint corner in English bond

(b) Squint corner in English bond

Figure 6.35 Obtuse angles in Flemish bond

An alternative method of building an obtuse angle is to overlap ordinary square bricks at the angle, called a birdsmouth. Figs 6.36 and 6.37 show this method being used in both English and Flemish bond.

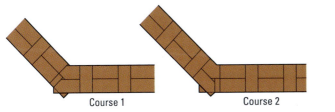

Figure 6.36 An alternative method of building an obtuse angle in English bond without using squint bricks

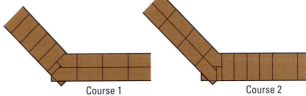

Figure 6.37 An alternative method of building an obtuse angle in Flemish bond without using squint bricks

The use of this technique for forming an obtuse angle saves the cost of the squint bricks, but great care is required to ensure accuracy in plumbing.

When the walling is built to form an internal obtuse angle, dogleg bricks are generally used to form the bonding at the angle, as can be seen in Fig 6.38.

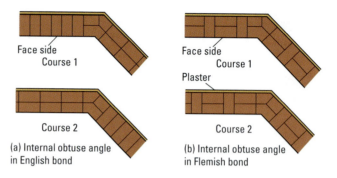

Figure 6.38 Internal obtuse angles in English and Flemish bond

This method is much stronger than with ordinary square bricks cut at the angle to form the bond, as it is then very difficult to ensure an adequate lap. Any poor lapping is likely to create a weakness and allow cracking to take place in the angle, should there be any slight movement in the structure.

PRACTICAL TASK

4. BUILD AN ACUTE (SQUINT) RETURN CORNER

OBJECTIVE

To practise building walls at angles other than 90°.

Squint quoins are special bricks which allow corners other than 90° to be formed and, as such, require different setting out methods. They can often be found on areas such as bay windows on houses.

This model uses 30 and 60° squint quoins to form the acute angled returns.

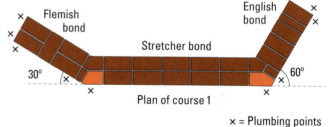

Figure 6.39 Using squint bricks to make angles

TOOLS AND EQUIPMENT

Walling trowel	Jointing iron
Lump hammer	Scutch hammer
Bolster chisel	Steel tape measure
Spirit level	Straight edge

PPE

Ensure you select PPE appropriate to the job and site where you are working. Refer to the PPE section of Chapter 1.

STEP 1 Mark a line long enough to locate the front line of the model as well as calculate the angles for the returns.

PRACTICAL TIP

The length of the whole bricks along the face will be 1,340 mm long and the extra for the added squints and angle marks can then be allowed for.

STEP 2 On the 60° angled return, mark a length of line a set distance, e.g. 400 mm. Then with a compass or trammel scribe lines from either end to a point where they intersect and form an equilateral triangle. The flank of this corner will be 60°.

PRACTICAL TIP

The calculation is based on the knowledge that all the internal angles in a triangle add up to 180°. Therefore, as all the angles are equal, when divided by three, the 60° angle will be formed when measured against the initial base line.

STEP 3 On the 30° angle, follow the same procedure as for the 60° angle. Once formed, mark a line half way alongside and draw a line through this and the corner opposite. This will bisect the 60° angle equally giving the 30° angle required.

STEP 4 Dry bond the first course making sure that all the angles and bonds are correct. The 60° return is constructed in English bond and the 30° in Flemish bond.

STEP 5 When you are satisfied with the setting out, lay the first course using level and straight edge to give a good base from which the rest of the model can be constructed.

STEP 6 Build up the corners on either end to the full height of six courses.

PRACTICAL TIP

Form the stop ends completely while forming returns as this is easier than racking them back and constructing after full height has been reached.

STEP 7 Run in the remaining brickwork on the face of the model.

STEP 8 Finish the model with a half round joint to the face and a flush joint to the rear.

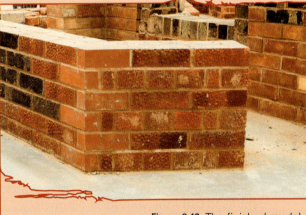

Figure 6.40 The finished model

TEST YOURSELF

1. Approximately how many standard size bricks will be required for a square metre of walling?

   a. 50

   b. 40

   c. 60

   d. 70

2. In rough arches, which part of the structure is wedge-shaped?

   a. Joints

   b. Bricks

   c. Plaster work

   d. Metal reinforcement

3. Which type of arch is non-structural and not in the shape of a circle?

   a. Flat

   b. Soldier

   c. Segmental

   d. Semi-circular

4. What material is usually used to make a turning piece?

   a. Concrete

   b. Metal

   c. Wood

   d. Polystyrene

5. Laggings can be fixed across the ribs of arch centres to carry the voussoirs. Normally these are small pieces of timber, but what is an alternative method?

   a. To use quick-drying cement

   b. To use metal rib reinforcement

   c. To use resin-bonded plywood

   d. There is no alternative

6. When building the corners of an arch, how many courses of bricks should be completed at any one time?

   a. 1 to 2

   b. 2 to 3

   c. 3 to 4

   d. 4 to 5

7. When building a semi-circular arch, which of the following is the last task?

   a. Remove supports and joint the soffit

   b. Allow the joints to contract as they set

   c. Lay the key brick

   d. Support the arch with props

8. What is a trammel?

   a. A type of brick course

   b. A wooden template

   c. A batten

   d. A piece of steel rod

9. Where are you likely to find a quoin?

   a. Under a wall opening

   b. At the corner of a wall

   c. As part of an arch

   d. As part of a chimney stack

10. Walls that are set at an angle of less than 90° are said to form which of the following?

    a. A right angle

    b. An acute angle

    c. A wide angle

    d. An obtuse angle

Chapter 7

Unit CSA–L3Occ134
CONSTRUCT DECORATIVE FEATURES

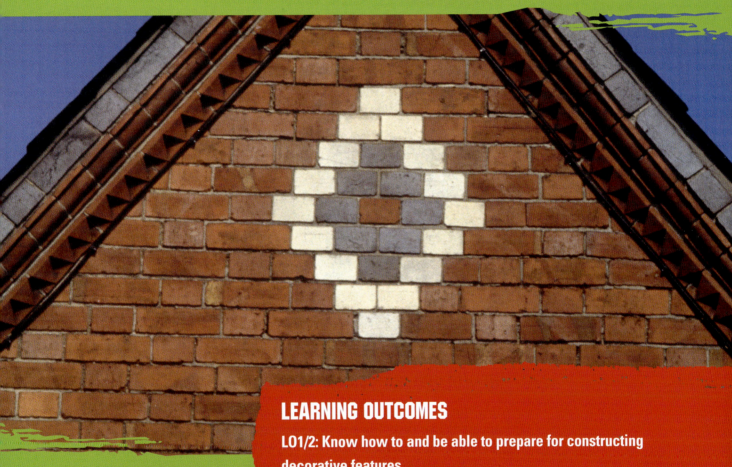

LEARNING OUTCOMES

LO1/2: Know how to and be able to prepare for constructing decorative features

LO3: Know how to set out and construct ramped brickwork

LO4/5: Know how to and be able to set out and construct battered walling to the given specification

LO6/7: Know how to and be able to set out and construct reinforced brickwork to the given specification

LO8/9: Know how to and be able to set out and construct brickwork incorporating features to the given specification

INTRODUCTION

The aims of this chapter are to help you to:

* select materials, components, tools and equipment

* construct decorative features.

PREPARING FOR CONSTRUCTING DECORATIVE FEATURES

Decorative bonds can be used for a wide range of features or panel work. These require considerable bricklaying skills, as the setting out procedure aims to ensure that any cut bricks are of the correct size for the panel or decorative feature.

Many architects will incorporate decorative panels into walling in order to provide features to make the building more attractive and interesting. The important thing, however, is to ensure that any of this decorative work is still capable of providing the essential support and stability required.

Decorative work can be used as a means of spanning openings. This was looked at in some detail in Chapter 6 when we looked at the construction of arches. Other decorative work is also incorporated into the construction of chimneys and chimney stacks, which were covered in Chapter 5. It is also worth remembering that many of the joint finishes which were also discussed in these chapters are relevant here.

Health and safety and hazards

It is important to remember that hazards can be encountered during any particular task. The construction industry does have its fair share of potential hazards. Each particular job should involve assessing potential risks. The risk assessments should highlight the precautions that are needed in order to prevent harm. If a hazard is significant then you need to take satisfactory precautions to reduce that risk. This means you do the following:

* Look for hazards.

* Decide who could be harmed and how.

* Evaluate the risks and decide whether any existing precautions are adequate.

* Record your findings.

* Review the assessment and revise it as necessary.

The process of looking at health and safety preventative measures is covered in some detail in Chapter 1.

Checking and interpreting drawings and specifications

We have looked at how to check that the drawings and specifications comply with standards. As many of these tasks related to constructing decorative features are detailed work, they could be affected by incomplete or incorrect drawings and specifications so you must check them carefully before you start work.

Reporting inaccuracies

If there are inconsistencies between working drawings and specifications then these should be highlighted before any construction work is carried out. The whole purpose of creating decorative features is to provide a focal point. If these focal points are incorrectly placed or constructed then the whole effect that was intended could be ruined.

While some of the decorative work may not be structural it is equally important to check that there are no inaccuracies before any work gets underway.

Potential hazards

Many of the potential hazards associated with the work methods and resources are relevant to general bricklaying work. These were looked at in Chapter 6 and you should also remember the potential problems which were outlined in Chapter 1.

The other consideration is that some of this type of work may well have to be carried out at height, so it is important to remember that all necessary precautions need to be taken to work from a secure platform. It is also important that anyone passing underneath the work is adequately protected with a hard hat. Ideally the area below should be a no-go zone while work is underway.

PPE

General personal protective equipment, including goggles or safety glasses and gloves, is essential, as with any bricklaying task. These should be worn at all times and without question. For other specialised work, such as working at height, harnesses or other equipment designed to prevent and arrest falls may be necessary. You may also need to wear a hard hat and safety boots.

Resources and calculating quantities

In Chapter 6 we looked at these issues in relation to complex masonry structures. Many of the resources that you will be using for decorative work will be the same as in normal building tasks. However, some of the decorative work does require the use of specially cut bricks or plinths, as well as the use of reinforcement methods.

As with all brickwork jobs, the area of the decorative work needs to be carefully measured out and a calculation made, based on the area, how many bricks or other resources will be necessary to complete the job.

It is always advisable to slightly over-order by around 5 per cent to take into account the fact that some of the bricks or other resources may arrive on site in an unfit state for use. Since you are intending to draw the eye to decorative work it is not acceptable to use damaged or faulty components, as these will be even more obvious than they would be in a normal brick wall.

Protecting the work and its surrounding area from damage

In Chapter 6 we looked at the general ways in which you should ensure that the area you are working on is protected from others and from the weather. You should also make sure that you do not contaminate or damage the surrounding area by your activities.

Some of the decorative work can mean working on a particular area for a longer period of time than would normally be required for standard bricklaying. It may also be necessary to wait for courses of bricks to harden off before you are able to proceed with the next stage of the task. The longer you are working on a particular area, the greater the probability that adverse weather conditions may impact on the work.

RAMPED BRICKWORK

A ramp is a decorative feature particularly used on boundary walls where a lower level wall joins a gate pillar.

To set out a straight ramp fix a line as shown in Fig 7.1.

The bricks should be marked with a bevel and then cut with a hammer and bolster (or mechanical saw). They should then be bedded to the line. The line then needs to be raised to bed each course of tiles and the brick on edge. The brick on edge should be mitred at the angles. The mitre should bisect the angle of the ramp, as can be seen in the following diagram.

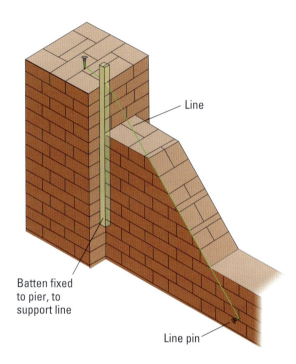

Figure 7.1 Method of cutting and preparing a ramp for a brick on edge ramp

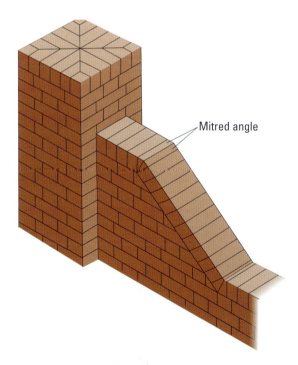

Figure 7.2 Brick on edge finish to a sloping ramp

The finished effect of the work can be seen in the following diagram.

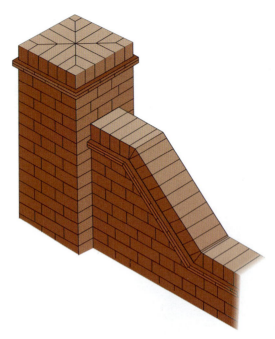

Figure 7.3 Brick on edge and tile creasing capping to a ramp

This is the method of building a circular ramp:

* Erect the pier to the height of the top of the ramp.

* Place a length of timber of a suitable size (75 × 50 mm would be adequate) that is long enough to extend beyond the striking point of the curve.

* Place dry bricks on top of this timber to weigh it down and keep it in place (see the following diagram).

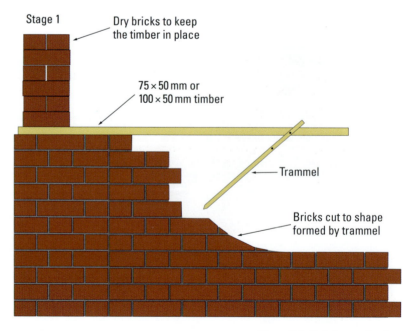

Figure 7.4 Building a circular ramp

● Fix a thin length of wood, 25 × 6 mm, equal in length to the radius plus about 400 mm, at the striking point of the curve of the ramp. This is called a trammel. The work is simplified if the trammel is sawn to a point at the free end, because the pencil marking of the bricks to be cut is kept on the centre line of the trammel, which makes the setting out as accurate as possible.

● Another method is to make the trammel longer and knock a nail through it at the appropriate position and use the head of the nail to scribe the curve onto the brick.

● Once the curve has been built, re-fix the trammel at a distance equal to the brick on edge plus one joint, and lay the brick on edge coping to the trammel (see the following two illustrations).

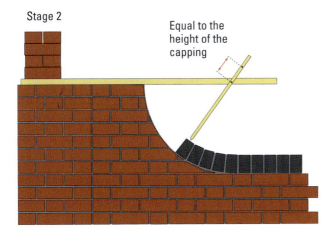

Figure 7.5 Completing the circular ramp

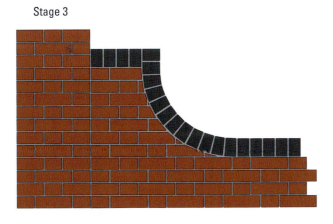

Figure 7.6 The brick on edge capping is complete

If the radius of the ramp is small, cut the bricks wedge-shaped in the same manner for cutting arch voussoirs, but if the radius is large then you may build the ramp with 'V' joints depending on the quality of the work required.

The following three practical tasks take you through the methods and procedures required for building concave, convex and sloping ramps.

1. BUILD A CONCAVE RAMP

OBJECTIVE

To build a concave ramp to match the adjacent drawing.

This model replicates a feature that is often used at the end of a run of piers or a decorative wall. They are often located next to a gate post, which increases their strength while allowing the height of the main body of the boundary wall to be reduced. Although this and the convex ramp look very complicated, careful setting out and positioning of a trammel will help to keep the process of construction fairly straightforward.

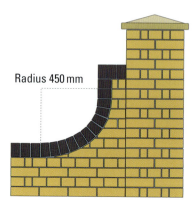

Radius 450 mm

Figure 7.7 Model of a concave ramp

PPE

Ensure you select PPE appropriate to the job and site where you are working. Refer to the PPE section of Chapter 1.

TOOLS AND EQUIPMENT

Walling trowel	Jointing iron
Lump hammer	Scutch hammer
Bolster chisel	Steel tape measure
Spirit level	Timber for trammels

STEP 1 Establish the position of the model and mark out on the floor.

STEP 2 Dry bond the first course and build the pier section. The projecting wall can be checked for alignment by placing a straight edge along the side of the pier and checking for equal measurement via a tape measure.

STEP 3 Construct the full wall up to the fourth course.

STEP 4 After the fourth course, construct the pier only up to the twelfth course, leaving indents for the ramp.

STEP 5 Weight a piece of timber in position on the top of the pier and fix the trammel. The trammel should be fixed to the centre point of the required radius and extend to the extrados of the curve.

Using the trammel point, mark off the required cut, making sure that both faces are accurate.

PRACTICAL TIP

Where a full width cut is required, it may be easier to cut the brick in two halves and use the trammel to mark the front and back faces of the wall so that both faces are kept as accurate as possible.

STEP 6 Bed the bricks into position, checking with the trammel, and ensure that there are no high spots that may interfere with the correct position of the capping bricks. Complete the cuttings to full height.

STEP 7 To bed the brick on edge capping, shorten the trammel by half a brick plus a 10mm joint allowance (112.5mm). The trammel can now be used to check the internal face of the curve.

PRACTICAL TIP

If the radius is gradual enough, the cappings can be laid with a 'V' joint as in a rough arch. If not, the bricks can be cut so that the final cut is level with the height of the ramp.

STEP 8 Finish the pier section to full height.

PRACTICAL TIP

If you build the brick on edge first you could bed up dropping mortar onto it when laying the rest of the pillar. It is best to build the pillar then put on the brick on edge.

STEP 9 Bed the capping stone, making sure that it is correctly placed and that it sits centrally on the pier.

STEP 10 Bed the three cappings to the top of the concave ramp and check for level in both directions.

STEP 11 Finish the main body of the wall with a half round joint. All cappings should have a flush joint to the top with a weather struck joint to either end.

PRACTICAL TASK

2. BUILD A CONVEX RAMP

OBJECTIVE

To build a convex ramp according to the plan.

As with the concave ramp, careful setting out and positioning of a trammel will help to keep the process of construction straightforward. This model could be built onto the previous model after taking down the concave section of ramp.

TOOLS AND EQUIPMENT

Walling trowel	Jointing iron
Lump hammer	Scutch hammer
Bolster chisel	Steel tape measure
Spirit level	Timber for trammels

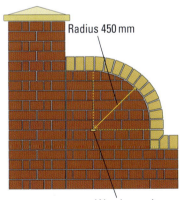

Radius 450 mm

Wooden pad

Figure 7.8 Model of a convex ramp

PPE

Ensure you select PPE appropriate to the job and site where you are working. Refer to the PPE section of Chapter 1.

STEP 1 Follow Steps 1 to 3 for Build a concave ramp, above.

STEP 2 After the fourth course, construct the pier only up to the twelfth course or remove the concave section of ramp, racking all the full bricks out, leaving only the cuts to be added later. Bed a piece of 10 mm timber into the ramp on either side so that the trammel can be fixed to mark the cut bricks and cappings on the ramp. This should be located at the point of the radius marked on the drawing.

STEP 3 Mark the trammel with the external edge of the cappings at its point, and a line marking the inner edge of the cuts 112 mm down from it.

PRACTICAL TIP

Where a full width cut is required, it may be easier to cut the brick in two halves and use the trammel to mark the front and back faces of the wall so that both faces are kept as accurate as possible. A trammel can be fixed on either side so that both sides of the brick can be marked at the same time.

STEP 4 Using the trammel, mark off the required cut, making sure that both faces are accurate.

STEP 5 Bed the bricks into position, checking with the trammel and ensure that there are no high spots that may interfere with the correct position of the capping bricks. Complete the cuttings to full height.

STEP 6 To bed the brick on edge capping, use the point of the trammels, ensuring that the bricks are laid horizontally and do not 'skew' to either side.

PRACTICAL TIP

If the radius is gradual enough, the cappings can be laid with a 'V' joint as in a rough arch. If not the bricks can be cut so that the final cut is level with the height of the ramp.

STEP 7 Finish the pier section to full height.

STEP 8 Finish with any remaining cappings to the ramp where it connects with the pier.

STEP 9 Bed the coping stone, making sure that it is correctly placed and that it sits equidistant on the pier in all directions.

STEP 10 Finish the main body of the wall with a half round joint. All cappings should have a flush joint to the top with a weather struck joint to either end.

PRACTICAL TASK

3. BUILD SOLID PIER WITH ATTACHED RAMP

OBJECTIVE

To build a solid pier with an attached ramp.

It demonstrates how a knee joint should be formed. Absolute accuracy is essential, and the cut bricks at the top and bottom of the rake need to be equal in order for the brickwork to be properly formed.

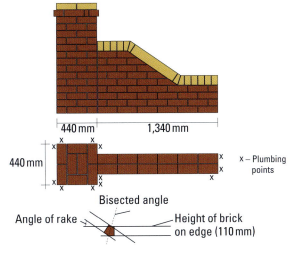

Figure 7.9 Model of a one-brick wall with a raking cut

TOOLS AND EQUIPMENT

Walling trowel	Jointing iron
Lump hammer	Scutch hammer
Bolster chisel	Steel tape measure
Spirit level	Timber for trammels

PPE

Ensure you select PPE appropriate to the job and site where you are working. Refer to the PPE section of Chapter 1.

STEP 5 Fix the line at the lower end of the rake by wrapping it around a brick or using another similar method.

STEP 6 Cut the bricks to the line using timber spacers to support the bricks to be cut. Mark the exact position of the line with a pencil, making sure it is not below or above it. This means that if it gets disrupted you can easily reset it to the correct position. Repeat for the other side of the ramp.

PRACTICAL TIP

Cut the wastage from the brick before cutting the proposed angle to the rake.

STEP 1 Mark the required position of the ramp on the floor, making sure it is 440 mm × 440 mm and equal across the diagonals. Ensure it is square to the pier by using a builder's square and straight edge.

STEP 2 Lay the entire first course and recheck for accuracy.

STEP 3 Build the pier and run out out the bricks to the ramp, whilst allowing for the bricks that have to be raked back.

STEP 4 At nine courses high, lay a piece of timber with a nail in the location of the top of the raking cut. Weigh the timber to stop the nail from moving.

STEP 7 Complete the remaining courses to the pier until the full height is reached.

STEP 8 Lay the oversailing courses to the top of the pier (25 mm all round).

PRACTICAL TIP

Make sure that the bed is raised on the outer edge to prevent the oversail bricks from leaning over.

STEP 9 Mark the position of the brick-on-edge capping and lay the bricks, ensuring that the joints are full in order to prevent damp penetration.

STEP 10 Apply the cement fillet by first adding mortar with a pointing trowel and then finishing with a consistent sweep of the entire length to prevent any lines in the finished result.

STEP 11 Lay the top and bottom horizontal bricks to the ramp so that the lines can be fixed to it and the bricks-on-edge can be accurately laid.

PRACTICAL TIP

Check that the bricks are positioned at the correct angle by using a boat level along the top arris.

STEP 12 Mark the bricks that are to be laid for the knee joints by determining the angle with an adjustable level. Mark on a piece of paper another set of lines to represent the top line of the brick-on-edge and then bisect the angle (see Fig 7.9). It is essential that both the cuts in the knee joint are equal and mirror each other. The angle for the top and bottom pair are the same and these marks can be used for both sets of cuts as long as they are reversed so that the faces are in the correct place on the top face of the ramp.

STEP 13 Attach the line and then mark the position of the bricks to be laid. If necessary, open or close up the joints so that the bricks are positioned equally along the rake. Then lay the bricks in along the length of the ramp.

STEP 14 Point all brick-on-edge cappings with a flush joint and a half round joint to the remainder of the model.

Figure 7.10 A wall with buttresses (1)

BATTERED WALLING

Any freestanding wall that is likely to take lateral pressure should be strengthened by introducing buttresses. Lateral forces can cause a wall to buckle. Buttresses can be attached to the wall at intervals to resist the lateral forces and give greater strength at the base of the wall. As less force is put on the wall nearer the top of the wall, the buttresses can be reduced in size as they get nearer the top of the wall.

The width of the buttress can be reduced either by battering the face of the wall or by using a stepped face, using plinth bricks, concrete cappings or tumbling in. See page 223 for more about tumbling in courses.

Ideally all retaining walls such as these, which are over 1 m high, should be battered back into the ground. In other words they should slope backwards around 10 to 15° in order to prevent collapse.

Brick, concrete and stone walling is built onto a reinforced concrete footing and the walls have weep holes. The need for retaining walls such as this is to deal with construction on a hillside or slope. The material is moved from the slope to form a level area. The material above that level needs to be held in place by the retaining wall. Therefore the wall needs to incorporate adequate drainage.

Setting out battered walling

All retaining walls need to be properly designed and constructed. Ideally a structural engineer should be involved.

There are a number of options, from reinforced concrete to the use of paving slabs. However brickwork can often be used for relatively low walls. Blockwork is quite economical for low to medium height walls, but these need considerable reinforcement and must be in-filled with concrete.

The first task is to establish the line of the wall. This can be achieved by driving a peg into the ground at each end of the wall. A string line can then be stretched tight from one to the other, representing the face of the wall line.

The next part of the setting out is to excavate a trench. The front end of this trench will act as a restraint for the lowest course of brick or blockwork. The trench should be filled with reinforced concrete and an even and level surface should then be achieved.

Construction methods and procedures

It will be necessary to ensure that there is sufficient drainage, so that water does not build up on the earth side of the walling. This can be achieved by laying pipes across the width of the wall to allow the water to drain away. Alternatively you can use weep holes.

In order to successfully achieve the correct angle of the retaining wall, batter boards are used. These are wedge-shaped boards that are flat on one side and slope on the other. They are fixed at each end of a section of wall and then the bricklayer's line can be attached to them at the edge of the wall. They can be used to check the inward slope of the brickwork. The batter boards need to be at right angles with the ground in order to give a true measurement.

Sloping piers or battered buttresses can be added in order to increase the lateral strength of the wall. The bed joints of the buttress are set at 90° to the slope. The following practical takes you through the steps in creating a battered wall and buttressed wall.

PRACTICAL TIP

Increasingly there are manufactured solutions to creating battered retaining walls. These are ideal for walls of up to 900 mm. These blocks are specifically shaped and use a locking pin to key the dry-laid walling system together. In effect these are designed to replicate traditional dry stone walling.

4. CONSTRUCT A BATTERED WALL AND BUTTRESS

OBJECTIVE

To build a freestanding wall with a buttress and battered wall.

Tumbling in is a decorative method of finishing off the top of a buttress. It allows the face side of the buttress to form a sloping side and provides a good surface to resist rain and frost. See page 223 for more about tumbling in courses.

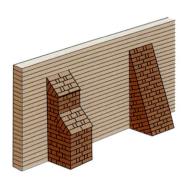

Figure 7.11 Wall with buttresses (2)

A gun template is used to maintain the correct angle of the slope.

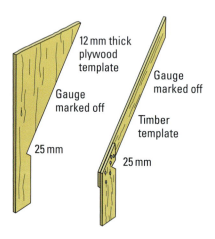

12 mm thick plywood template

Gauge marked off

Gauge marked off

Timber template

25 mm

25 mm

Figure 7.12 Gun template

A freestanding wall can be designed to have a battered (sloping) face; a special battered plumb rule or template is required to erect the battered face of the wall.

SPECIFICATION

* Set out the buttress and construct gun template before starting.

* Batterboard is to be cut or an adjustable level is to be set to the correct angle.

* Battered work to be weather struck, the end piers to be half-round, and the rear to be left flush.

* Risk assessment to be completed before work commences.

TOOLS AND EQUIPMENT

Walling trowel	Builder's square
Pointing trowel	Tape measure
Spirit level	Gun template
Lump hammer and bolster chisel	Battered plumb rule
	Adjustable bevel

PPE

Ensure you select PPE appropriate to the job and site where you are working. Refer to the PPE section of Chapter 1.

Figure 7.13 Using a battered plumb rule

In this practical task you are asked to build the freestanding wall with a buttress and battered wall as shown in Fig 7.14.

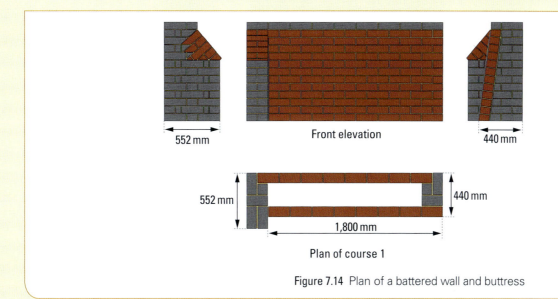

552 mm

Front elevation

440 mm

552 mm

440 mm

1,800 mm

Plan of course 1

Figure 7.14 Plan of a battered wall and buttress

STEP 1 Set out and dry bond the first course to the correct length, as shown on the drawing. Check the length of the wall.

At this point ignore the battered wall on the face but allow for it to be built on completion of the main wall.

STEP 2 Lay three courses of bricks at each end of the wall for the piers as shown on the drawing. Don't forget to cut the bricks at the smaller pier to allow for the slope of the battered wall.

STEP 3 Now lay the blocks in between the piers for the first course.

STEP 4 Continue until the wall reaches the eighth course.

STEP 5 At this stage it is better to lay the first course of the battered wall and allow it to harden before continuing to build it.

Once you have laid the first brick, check that it is at the correct angle by using the battered plumb rule and a spirit level.

PRACTICAL TIP

You will have to cut off part of the frog on one side of the brick to allow you to bed the brick down at the correct angle.

STEP 6 You now need to make the gun template for the tumbling in:

• The tumbling in is set out full size on a board to determine the size and shape of the cut bricks.

• Use the timber gun template to maintain the required slope. You can mark out the gauge onto the gun template.

• Every time the gun template is used, press the stem firmly against the plumb face of the attached pier, checking the slope of the face and that the tumbling in course is at right angles to the template, using a square.

STEP 7 Using the gun template, set an adjustable bevel to the correct angle for the tumbling in.

Measure the length of the first brick and cut it to the correct angle.

Build the first course of the tumbling in. Check that the brick is 90° to the gun template by using a small square.

STEP 10 Continue to build the tumbling in and the rest of the model until the twelfth course has been reached. Mark out, cutting the bricks in the same manner as in Step 7.

STEP 8 Cut the other two courses in the same manner.

STEP 11 Continue to build the rest of the battered wall using the battered plumb rule and spirit level to plumb the wall.

STEP 9 Lay the ninth course, bonding it into the tumbling in.

STEP 12 Once you have reached the top, finish the wall with a one-and-a-half brick thick course.

REINFORCED BRICKWORK

Under normal circumstances the strength of brickwork is adequate for carrying loads that bear directly down onto the walls. These loads create pressures which are trying to crush the wall. This is what is meant by the term 'state of compression.' Brickwork, however, is comparatively weak in resisting loads which create a stretching force, which is known as a 'state of tension'. It is for this reason that brickwork is never built directly across window or door openings unless a steel reinforcement is introduced so that the tensile forces can be resisted by the steel and not the wall itself.

There are other instances when tension can be caused in walling, by loads which act on the side or face of the wall, such as a pillar supporting a heavy gate (see Fig 7.15).

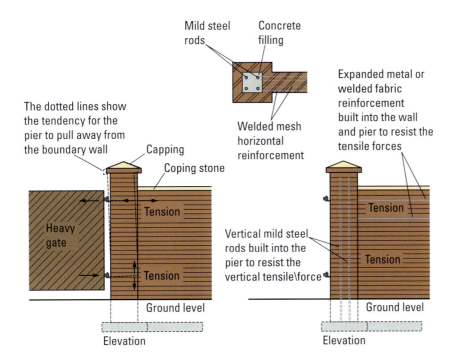

Mild steel rods

Concrete filling

Expanded metal or welded fabric reinforcement built into the wall and pier to resist the tensile forces

Welded mesh horizontal reinforcement

The dotted lines show the tendency for the pier to pull away from the boundary wall

Capping

Coping stone

Tension

Heavy gate

Vertical mild steel rods built into the pier to resist the vertical tensile\force

Tension

Tension

Tension

Tension

Ground level

Ground level

Elevation

Elevation

Figure 7.15 Reinforced brickwork for a gate pillar

Setting out reinforced brickwork

The longer and higher an exposed wall is built, the more difficult it will be to resist the side thrusts of the forces acting upon it. Unless some form of reinforcement is introduced into the brickwork, it is liable to fail. Reinforcement increases the lateral strength of the walling and helps to prevent failures.

Reinforcement can also be used where a decorative panel is introduced into a wall and a series of straight joints have to be used. In this case reinforcement may be built into the joints to bond the wall together and to prevent failure in the decorative panel. There are several different types of reinforcement, which are described in the following table and set of illustrations.

Type of reinforcement	Description
Expanded metal	This is a perforated metal strip. The holes help form a good key between the reinforcement and the mortar. It is supplied in coils of various widths suiting a variety of different types of walling.
Welded fabric	This consists of a pair of rods with connecting rods welded to them at intervals. The structure of the rods allows a good key to be created between the mortar and the steel. The welded fabric is supplied in rolls and it is good for building into walls which are to be carried over openings.
Mild steel rods	These provide a good method of reinforcing brickwork where vertical reinforcement is needed. They are good for reinforcing brick lintels, gate pillars or in the use of Quetta bonds. Reinforced walling can also be constructed using hollow concrete bricks which are laid at a half bond with the reinforcement is placed in the hollows. The hollows are filled with concrete as the wall is being built.

Table 7.1 Different types of reinforcement

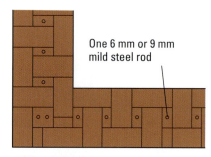

One 6 mm or 9 mm
mild steel rod

Plan showing alternative method
of placing reinforcement

Figure 7.16 Reinforced brickwork

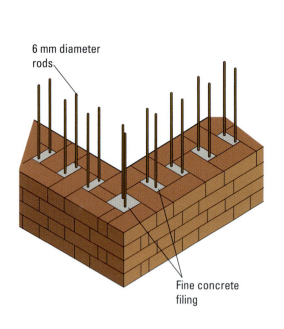

6 mm diameter
rods

Fine concrete
filing

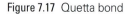

Figure 7.17 Quetta bond

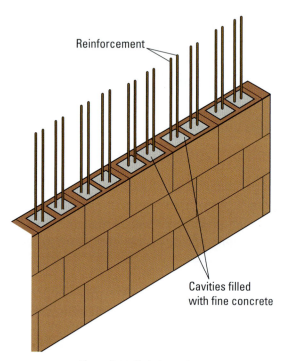

Reinforcement

Cavities filled
with fine concrete

Figure 7.18 Reinforced concrete block walling

Construction methods and procedures

Panels of brickwork can be reinforced in order to increase their
strength. Reinforcement can also be used at particular points of a
wall to resist the chance of cracking, such as at corners of openings.
They can also be used over openings, essentially to form a lintel in
the brickwork. The big advantage is that a clean brick profile is visible
without a separate steel lintel.

Special provision has to be made to support the lowest course of bricks, which will be below the reinforcement. The simplest type of reinforcement for brick walls is to use wire reinforcement in the bed joints. This tends to be done above and below openings. It is the corners of openings that create the stress and have the greatest chance of cracking.

The most common form of reinforcement is to use what looks like a ladder of wires. They have 3 to 6 mm longitudinal wires with thinner wires welded across them. The reinforcement has to be completely enclosed by the mortar.

One-and-a-half brick walls laid in Flemish bond will usually have cut bricks. These can be used as the cores into which the vertical bars can be fixed. They can be grouted in after several courses have been laid. There are also bricks with purpose-made core holes available. These core holes need to be located in the bricks so that they will align in a normal stretcher bond.

PRACTICAL TASK

5. REINFORCE A WALL WITH QUETTA BOND

OBJECTIVE

To build a one-and-a-half brick wall with reinforcement.

When walls need to be reinforced for strength then Quetta bond is used on walls that are one-and-a-half bricks thick. The wall shows Flemish bond on the face but the section through it is different as you do not fill in the voids in the centre of the wall. This leaves quarter-brick pockets where vertical steel reinforcement can be placed and the void filled with concrete.

The vertical steel bars are inserted into the concrete foundation and the wall built around them.

TOOLS AND EQUIPMENT

Walling trowel

Pointing trowel

Spirit level

Lump hammer and bolster chisel

Builder's square

Tape measure

Bat gauge

SPECIFICATION

* The wall is to be built four courses high.

* Reinforcing bars are to be installed and encased from course two.

* Joints are to be left flush as the work proceeds on both faces.

PPE

Ensure you select PPE appropriate to the job and site where you are working. Refer to the PPE section of Chapter 1.

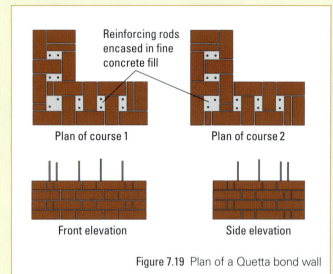

Reinforcing rods encased in fine concrete fill

Plan of course 1   Plan of course 2

Front elevation   Side elevation

Figure 7.19 Plan of a Quetta bond wall

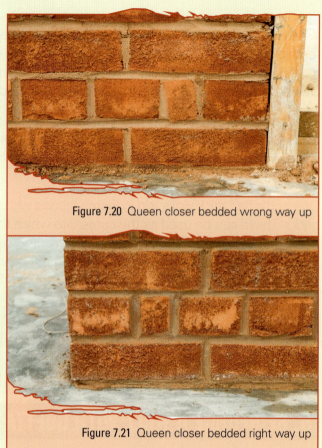

Figure 7.20 Queen closer bedded wrong way up

Figure 7.21 Queen closer bedded right way up

STEP 1 Set out and dry bond the first course to the correct length as shown on the drawing.

STEP 2 Cut the correct number of queen closers to the correct size using the appropriate tools and PPE. Some bricklayers prefer to cut the bricks on old carpet or another soft surface to absorb the impact of the cut.

PRACTICAL TIP

Queen closers are cut down the length of a brick. It is important to cut the closers to the correct size. If they are too big then it will throw out the bond on the header course and subsequently loose the bond and you will end up with straight joints. Do not just cut the brick down the middle as this will be too wide and will make it difficult to keep the bond correct.

Always lay the queen closers with the full face of the bed of the brick down. If you lay the closers 'frog' down you only have three edges and the closer will tend to tilt and this will show up in the wall.

STEP 3 Set out a 90° angle using a builder's square and check it for accuracy using the 3:4:5 method.

STEP 4 Lay the corner brick on the face of the wall to gauge and level both ways. Lay the remaining 4 bricks along the face, starting with a header, levelling them with the corner brick.

STEP 5 Now lay the bricks on the return corner. Check that the corner is 90°.

STEP 6 Lay the second course and leave the mortar joints to harden before inserting any concrete in the voids.

PRACTICAL TIP

If you start filling the voids with concrete at this stage you will only push the bricks out of place if the joints are still wet.

STEP 7 Once the mortar joints have sufficiently hardened, carefully fill the voids with a weak concrete. Remember not to force the concrete into the void too hard otherwise you will only move the brickwork.

STEP 9 Continue to build the rest of the wall until it reaches the correct height.

STEP 8 Insert the reinforcing bars into the concrete and make sure that they are plumb.

CASE STUDY

Choosing the right candidates

David Williams, Regional Labour and Apprentice Manager at Laing O'Rourke, talks about the application process and apprentice programme.

'I'm responsible for coordinating all labour onto a site, which trades they need, and I make them available through sub-contractors, employees and apprentices. When I joined the company nine years ago, they wanted to become known for their apprenticeship programme. Now, we run several different kinds of programme, such as our "Apprenticeship Plus" that takes four years to complete and opens up the door to higher achievers who we hope will become our future managers and supervisors.

The application process for all our apprenticeships starts with an online form. The applications are sifted, looking for people with potential and a good basic education. We generally select applicants with GCSEs in Maths, English and Science with a grading of A–C.

We then conduct a phone interview, which gives a good indication of a person – you can tell a lot about someone's attitude over the phone. A group of people will discuss and decide which applicants will then be invited to attend an assessment morning in their region.

They do a one-to-one interview, some fun team building exercises (to help people relax and be themselves), and a dexterity test that looks at hand–eye coordination. The Apprenticeship Plus applicants also have to make a 10-minute presentation. There's no one right answer or approach to any of these activities; it's to bring out their individual characters and skills. Someone might do poorly in one area, but perform excellently in another and therefore get through. It's not just about what you can do academically – showing a positive attitude is a major factor, as well as how you present yourself. If you turn up to the interview and present yourself in a haphazard way, then we'll wonder what you'll be like every day.'

BRICKWORK INCORPORATING FEATURES

Although the main purpose of bonding is to stabilise and strengthen brickwork, some bonds are intended for decorative purposes only. The decorative work gives relief and depth in appearance by creating shadow lines. They form a design which may otherwise appear to be just a mass of brickwork. The decorative work does add extra cost, but it is possible to introduce some features at a fairly economical price.

Some of the more economical methods include:

* the use of contrasting coloured pointing mortars

* introducing different coloured bricks into the walling

* laying bricks in special patterns

* laying bricks that project from the face of the work, creating shadow lines on the face of the walling.

Setting out and constructing brickwork incorporating flush and projecting features

There are several different ways in which features can be incorporated into brickwork. Each of these sections details the setting out, construction and procedures as appropriate.

Projecting bricks

These are used in many structures to form patterns on what would otherwise be large, plain surfaces of brickwork. The simplest method is to use headers that project from the face of the wall by about 18 mm. These are laid at equal distances apart and kept plumb so that straight lines are formed horizontally, vertically and diagonally across the wall. A small template or gauge can be made in order to help check the accuracy of the projections. An example can be seen in Fig 7.22.

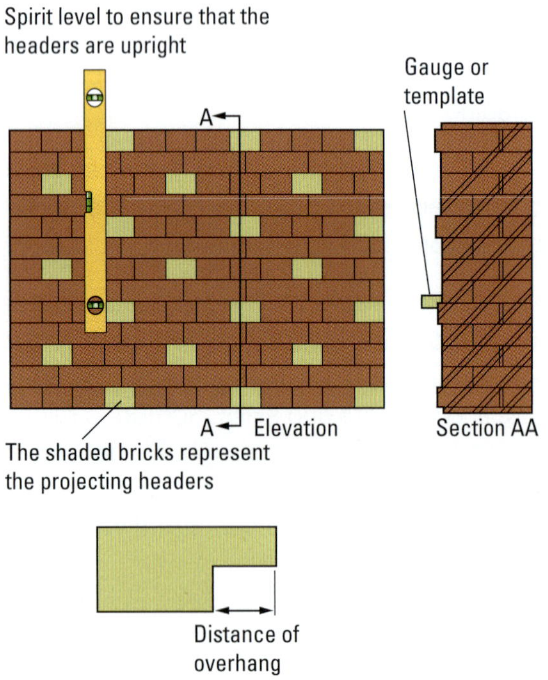

Figure 7.22 Using a plywood gauge or template when building a wall with projecting bricks

Another method is to cut the ends of the bricks so that a rough surface is left protruding from the face of the wall. Although the surface is rough it is important to take care with the cutting because it is essential that the edges are cut square while the surface is left uneven, as can be seen in the following diagram.

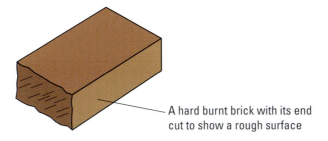

A hard burnt brick with its end cut to show a rough surface

Figure 7.23 Projecting bricks for decorative brickwork

Another method is to use small isolated patterns in the wall. They can be effective if contrasting coloured bricks are used for the patterns.

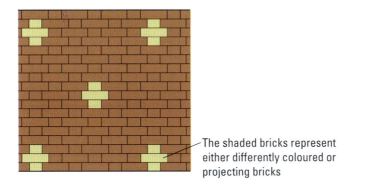

The shaded bricks represent either differently coloured or projecting bricks

Figure 7.24 Decorative brickwork with contrasting coloured and projecting bricks

Diaper bond

On large areas of plain walling diamond patterns can be formed using coloured bricks which are in contrast to the general walling. This is a popular way of creating a feature in brickwork and the process can be seen in the diagrams below.

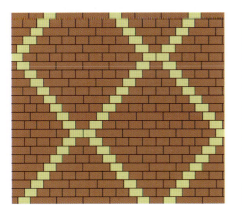

Figure 7.25 Diaper bond (1)

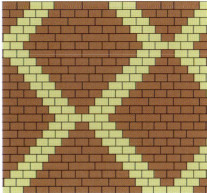

Figure 7.26 Diaper bond (2)

Basket weave bond

Basket weave bonds are mainly used in panels. They consist of sets of three bricks which are laid alternately horizontally and vertically. Sets of four bricks can also be used, providing the four bed joints are equal to the length of one brick. Examples can be seen in the following two diagrams.

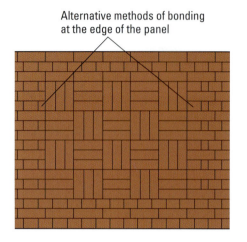

Figure 7.27 Basket weave bond using three courses to 225 mm

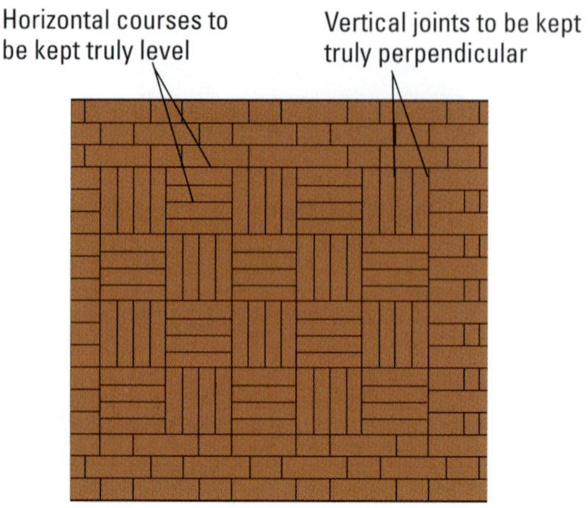

Figure 7.28 Basket weave bond using four courses to 225 mm

The important points to watch when building panels with this bond are:

* keep the vertical bricks truly plumb and the horizontal courses level

* select bricks carefully for consistency in length so that the patterns are of even shape

* keep the main horizontal and vertical joints that pass through the panel in straight lines.

Although this appears to be a quite straightforward type of bond, a great deal of skill and care is essential. A variation of this bond is building the basket weave pattern at 45° to the horizontal. This can produce a very effective panel, but it involves a great deal of cutting at the edges, as can be seen in the following diagrams.

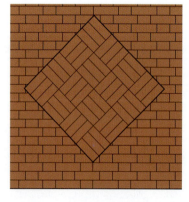

Figure 7.29 Diagonal basket weave bond (1)

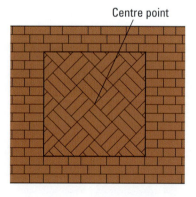

Figure 7.30 Diagonal basket weave bond (2)

Whenever the panels in a diagonal basket weave bond are being built it is good practice to set out the panel first on a suitable flat surface. Bricklayers will use hardboard, plywood or even a concrete floor or slab. The bricks are then carefully marked out and cut to their required shapes. All this is done before building them into the panel.

Another way to set out these types of panels is to lay the bricks dry on a flat surface which takes into account the correct width of joint between the bricks. With the aid of a straight edge the outline of the panel is marked out on the top of the bricks, as can be seen in the following diagram.

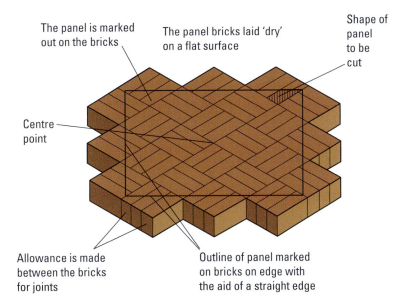

Figure 7.31 A method of setting out a panel in diagonal basket weave bond

Another method is using a temporary timber frame constructed to the right size. This can be laid over the top of the bricks and the edges of the panel can then be accurately marked out.

The bricks that are on the edge of the panel are now marked ready for cutting and this can be achieved by either using a hammer and bolster or a mechanical saw.

Herringbone bond

This consists of a series of patterns of bricks that are laid out at 90° to each other but at 45° to the horizontal plane. The patterns, as can be seen in the following two diagrams, can be laid out vertically or horizontally.

Portion of brick
required

Stage 1

The brick is marked out
for a cut in a herringbone
pattern

Stage 2

The brick is cut with a
hammer and bolster at
right angles to its face

This surface trimmed
with a comb hammer

Stage 3

The brick is cut along
the splayed line and the face
of the cut trimmed to an even
surface

Figure 7.34 Cutting bricks with a
bolster and hammer

The patterns must be
kept vertical

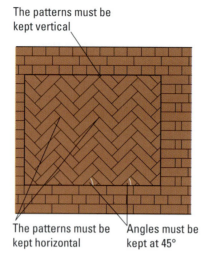

The patterns must be
kept horizontal

Angles must be
kept at 45°

Figure 7.32 Vertical herringbone bond

Figure 7.33 Horizontal herringbone
bond

This work should be carried out by either of the methods described above for the diagonal basket weave bond. When panels are being built with herringbone bonds, care needs to be taken to make sure that each pattern is kept truly vertical or horizontal. This means checking all of the angles in the pattern and that they are maintained in a straight line.

When cutting the bricks with a hammer and bolster, you should not attempt to cut the brick directly along the slanting line. It will only break off at an acute angle and ruin the brick. You should cut the brick at right angles along its thickness from the point where the slanting line meets the edge of the brick. Then you should cut along the slanting line and this will produce a clean-cut brick, as can be seen in Fig 7.34.

Diagonal herringbone bond

This can be created by moving the herringbone patterns to 45° to the horizontal. This provides a good-looking pattern but with the advantage that it requires far less cutting than either vertical or horizontal herringbone bonds. An example can be seen in Fig 7.35.

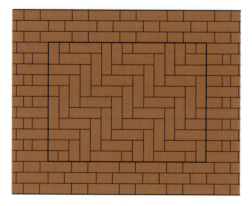

Figure 7.35 Diagonal herringbone bond

Double herringbone bond

These are very similar to herringbone bonds. The main difference is that double stretchers are used instead of single bricks. This can be seen in the following two diagrams.

Figure 7.36 Double herringbone bond

Figure 7.37 Diagonal double herringbone bond

All these bonds are suitable for the building of decorative panels and infilling the walling between a lintel over an opening and the relieving arch over the lintel, known as a tympanum. An example is shown in the following diagram. Note that these are rarely used in modern construction.

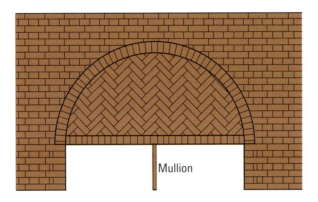

Mullion

Figure 7.38 Filling in a tympanum with herringbone bond

Tumbling in courses

These are used where buttresses or walls have to be reduced in size. They allow the face sides of the bricks to form the sloping side of the work and provide a good surface to resist rain and frost. In Fig 7.39 tumbling in is shown making use of plinth bricks.

Fig 7.40 shows the laying of bricks at an angle with their beds cut to suit the angle of the slope of the tumbling in.

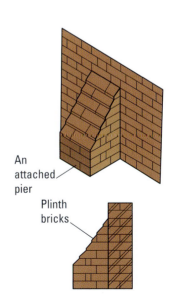

An attached pier

Plinth bricks

Figure 7.39 Tumbling in with the aid of plinth bricks

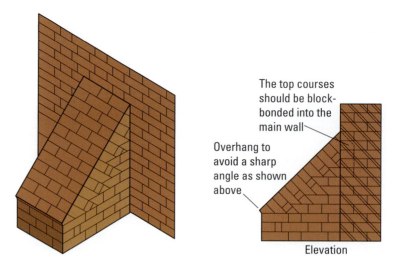

The top courses should be block-bonded into the main wall

Overhang to avoid a sharp angle as shown above

Elevation

Figure 7.40 Tumbling in courses

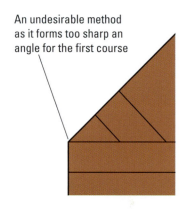

An undesirable method as it forms too sharp an angle for the first course

Figure 7.41 Undesirable method of cutting the first course of tumbling in

It is usual to overhang the first course of the tumbling in to prevent the first bricks from having to be cut at a sharp angle and only leaving a small area of brick that can be affected by the weather. Fig 7.41 shows the problems with this.

The following methods are used to ensure the accuracy of the work when building tumbling in courses:

* One batten is fixed to the wall so that it cuts the point where the top of the tumbling in will occur and another at the bottom of the tumbling in. This enables you to fix a line from the top to the first course of the work and the tumbling in course can now be built in to the line.

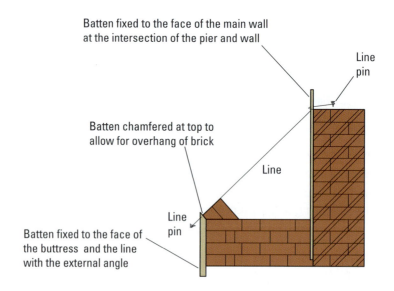

Batten fixed to the face of the main wall at the intersection of the pier and wall

Line pin

Batten chamfered at top to allow for overhang of brick

Line

Line pin

Batten fixed to the face of the buttress and the line with the external angle

Figure 7.42 Method of lining in tumbling in

* A gun or template is cut from hardboard or plywood.

* Two pieces of batten are fixed together to form a gun to suit the angle of the slope of the tumbling in courses. These may be used to check the accuracy of the work by placing the template or gun at the right height and against the work immediately below the tumbling in.

* The distance from the bottom to the top of the tumbling in should be gauged to prevent the use of a split course, which would look ugly.

If the tumbling in is only small all of the tumbling courses may be taken down to the same horizontal course.

If the work is quite extensive and will require a large number of courses, the tumbling in can be divided into sections. This gives a very dramatic effect. The sections should be kept in similar sizes and shapes so that there is a reasonable balance between the horizontal and tumbling courses, as can be seen in the following diagram.

PRACTICAL TIP

It is important that this work is kept truly plumb. Before beginning to build the tumbling in courses, the work should be carefully set out on a board or other suitable flat surface.

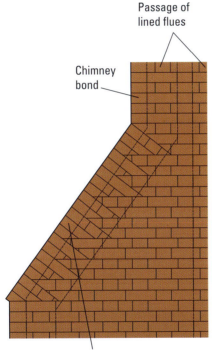

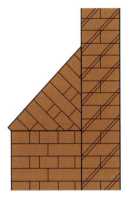

Figure 7.43 Typical method of bonding for a small tumbling in

Three quarter batts have been used in the tumbling in courses to match the chimney bond but closers could have been used next to the header

Figure 7.44 Tumbling in courses for a chimney stack

Gable shoulders and springers

The following illustrations show a wide range of different designs used at the springing of gables. They can be constructed out of brick, tiles or concrete.

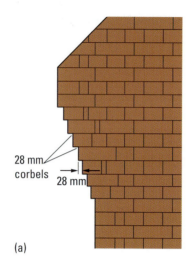

(a)

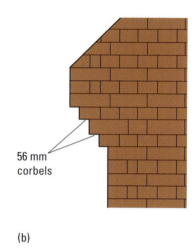

(b)

Figure 7.45 Examples of brick gable shoulders

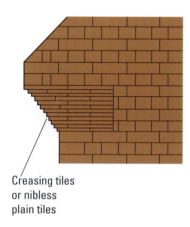

Figure 7.46 A tiled gable springer or tiled knee

Figure 7.47 Ornamental tiled gable springer

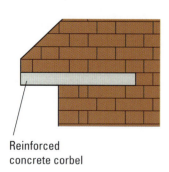

Figure 7.48 Reinforced concrete gable springer

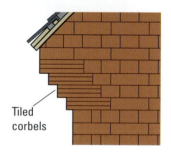

Figure 7.49 Gable tile springer

If bricks are used the bond needs to be adjusted to suit the overhanging courses. It is normally considered to be safe practice for the overhang not to exceed the thickness of the wall immediately below the overhang.

String courses

String courses are horizontal courses that are built into the face of walls to form an architectural feature. These can include the following:

* Soldier courses – these are bricks that are laid on end side by side. They have to be laid truly plumb otherwise the bricks will give the appearance that they are leaning over.

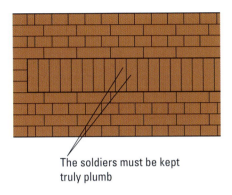

The soldiers must be kept truly plumb

Figure 7.50 A soldier string course

* Moulded bricks – these can project from the face of the wall. The projecting courses need to be lined up on the bottom edge of the bricks rather than the upper edge. This will mean that any irregularities in the thickness of the bricks are taken up on their upper edges. The eye line at the lower edge remains straight. Figs 7.51 and 7.52 show this and an example of the types of mouldings that are available. If the string course is above your eye line, then the lower edge must be lined in and level. If it is below your eye line then the top edge must be lined in and level.

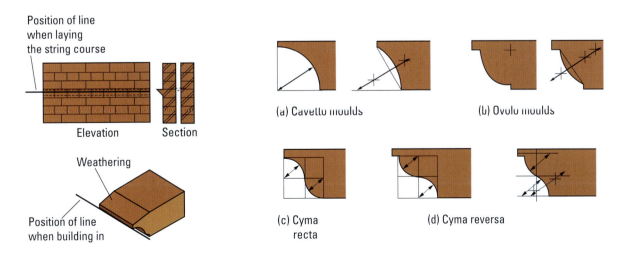

Position of line when laying the string course

Elevation   Section

Weathering

Position of line when building in

(a) Cavetto moulds

(b) Ovolo moulds

(c) Cyma recta

(d) Cyma reversa

Figure 7.51 Details of a typical moulded brick

Figure 7.52 Different types of mouldings showing methods of their setting out

Dentil courses

These are designed to provide a decorative feature on the upper surface of a wall. They can often be seen at eaves level. They are formed by using projecting bricks in patterns, as can be seen in the following diagrams.

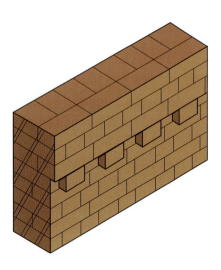

Figure 7.53 A single dentil course

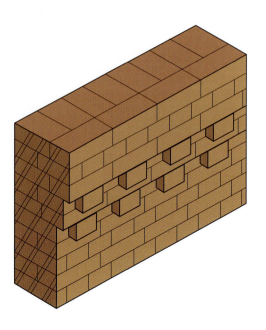

Figure 7.54 A double dentil course

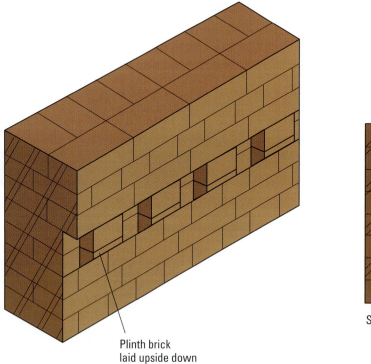

Plinth brick
laid upside down

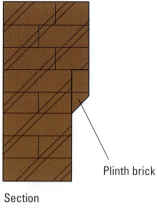

Plinth brick

Section

Figure 7.55 A single dentil course incorporating the use of plinth bricks

Dog toothing

These are laid in a similar way to dentil courses but they are laid at 45° to the face of the wall on plan. They can be recessed or projecting. They can also be laid as a single course or in multicourse panels and string courses, as can be seen in Fig 7.56.

Over-sailing courses

In Chapter 5 we saw that over-sailing courses are often used as a decorative feature in chimney stacks.

The term over-sailing can refer to any course of brick that projects out from the face of the wall. It is also often referred to as corbelling. Each successive course of brickwork overhangs the previous one, which in effect moves the building line. The over-sailing should always be just less than half a brick. This will ensure that once the structure is loaded then it will have good strength. In practice if you are using a brick as a header the over-sail is kept to a maximum of 50 mm. It will be half that if the bricks are laid as stretchers.

As with all types of decorative work it is important that the following points are taken into account:

- Care needs to be taken to ensure that all work is truly plumb and level.

- All overhanging work needs to be lined in on the bottom edge of each course.

- Work involving cutting should be carefully set out before construction begins.

- Always use templates, trammels, guns and gauges to check the accuracy of the work.

- Carefully select bricks to be used on these special features and always reject any that have damaged faces.

- Always check the angles at which bricks are being cut and laid.

Joint finishes

Interesting patterns can be created in face work using coloured mortars for pointing. One method is to blind out all of the perpends with a mortar that matches the colour of the face bricks. The bed joints are then pointed with a light or contrasting coloured mortar. This gives the visual effect of the wall being built with long, thin slabs.

A second method is to use a flying Flemish bond and then blind out the joint between each pair of stretchers with a mortar matching the colour of the bricks. The remainder of the pointing is finished using a contrasting coloured mortar. The stretchers in each case appear to be elongated, as can be seen in Fig 7.57.

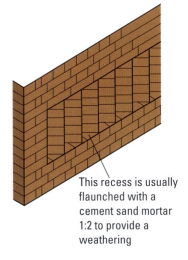

This recess is usually flaunched with a cement sand mortar 1:2 to provide a weathering

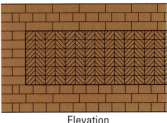

Elevation

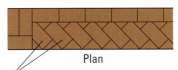

Plan

The bricks in the dog toothings are laid in opposite directions in alternate courses to form a bond

Figure 7.56 Dog toothing

REED TIP

A common interview question is 'Tell us about yourself'. It sounds easy, doesn't it? But take advantage of this question and use it to talk about the things you do that will help you be a good tradesperson, e.g. weight training at the gym or working with the community.

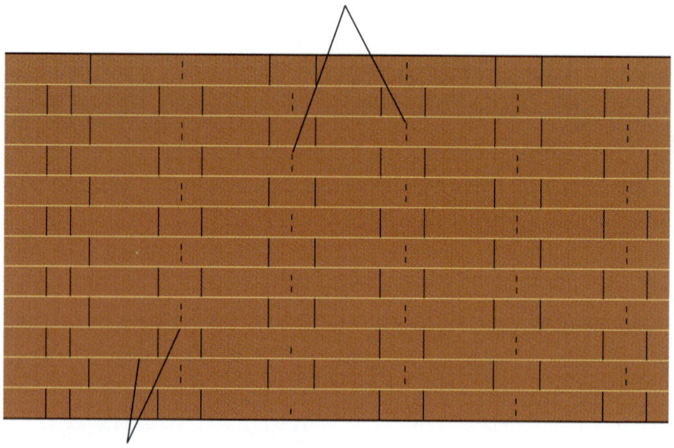

The perpends between stretchers
'blinded out' with a mortar of the
same colour as the bricks

The bed joints pointed with a
contrasting coloured mortar

Figure 7.57 An elevation showing the use of coloured mortars for pointing

A third method is to have two colours; one for the perpends and the other for the bed joints. This can greatly enhance the colour of the bricks. There are many different colouring agents that are available for use in mortars, but it is important to remember when mixing them that:

* the mix must be carefully gauged and the same gauge is used for each mix

* the same class and type of sand is used for each mix.

In Chapters 5 and 6 we looked at different types of jointing. A recessed joint should only ever be used with frost-resistant bricks and in fairly sheltered exposure conditions. They can provide a good articulated joint.

The alternative is to consider a weather struck joint. These have excellent strength and weather resistance and they are suitable for all types of exposure to weather.

The difference between these joints is that a recessed joint has a squared off recess of up to 3 to 4 mm while the weather struck joint has a downward sloping angle to encourage water to run away from the joint and down the face of the brickwork.

The following practical tasks take you through several of the different types of decorative brickwork step-by-step, including their setting out.

PRACTICAL TASK

6. COMPLETE A SECTION OF STRAIGHT FLEMISH BOND WALL INCLUDING DIAPER WORK

On large areas of plain walling diamond patterns can be formed using coloured bricks which contrast with the general walling. This is a popular way of creating a feature in brickwork. The contrasting bricks can be projecting, recessed or flush with the main wall.

OBJECTIVE

To complete a section of a Flemish bond wall including a decorative pattern.

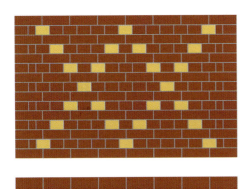

Figure 7.58 Flemish bond wall with diamond (diaper) pattern

The model is to be built in local facing bricks with contrasting headers to form the diamond pattern. These will project 15 mm.

Joint finish:

Front – half round

Rear – flush from the trowel

TOOLS AND EQUIPMENT

Walling trowel

Pointing trowel

Spirit level

Lump hammer and bolster chisel

Builder's square

Jointing iron

Corner blocks and line

Batt gauge

Tape measure

PPE

Ensure you select PPE appropriate to the job and site where you are working. Refer to the PPE section of Chapter 1.

STEP 1 Set out and dry bond the first course.

STEP 2 Cut the queen closers to the correct size using the appropriate tools and PPE.

STEP 3 Lay the first brick at one end to gauge and level. Now lay the second end brick. Using a straight edge and level, level the second end brick with the first.

STEP 4 Using corner blocks and line run in the first course.

STEP 5 Build two small corners (five courses) at each end. Make sure that you keep the bond. Don't forget the contrasting header on the second course. Make sure that it is projecting 15 mm and that the brick has no visible defects on the header face that will be seen.

STEP 6 Using corner blocks and line run in the courses between the corners.

PRACTICAL TIP

It is important to have the contrasting bricks in the right place to maintain both the bond and the right pattern throughout the wall; therefore it is best to lay the contrasting bricks in position and wall the rest of the course up to them. Then you can bed the contrasting bricks in position.

STEP 7 Repeat Steps 5 and 6 until you have reached the correct height. Keep checking the contrasting bricks are projecting 15 mm.

PRACTICAL TIP

Make sure that you plumb the contrasting bricks not only on the face but up one side as well.

PRACTICAL TASK

7. CONSTRUCT A BASKET WEAVE PANEL

OBJECTIVE

To set out and construct a decorative basket weave bond panel.

The arrangement of the bricks in this model is for purely decorative purposes. As with the herringbone panel, there is a long tradition of this type of work being used. It was often incorporated between the spaces on Tudor timber frame houses when the wattle and daub infill was released and as bricks became more popular. The use of any decorative work often indicated wealth, as the time spent on such work showed that the customer could afford to pay the bricklayer to spend time on more intricate work.

For the two decorative panels, it is recommended that a section of 'dummy' wall is built to contain them, comprising a wall built with blocks laid flat and with the appropriate sized opening.

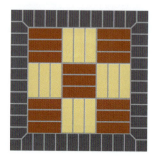

Figure 7.59 Plan of vertical basket weave

PPE

Ensure you select PPE appropriate to the job and site where you are working. Refer to the PPE section of Chapter 1.

TOOLS AND EQUIPMENT

Brick trowel

Lump hammer

Bolster chisel

Spirit level

Jointing iron

Scutch hammer

Steel tape measure

Lines

PRACTICAL TIP

Before starting to build the panel it is best to select all your bricks and make sure that they are all the same size and have no defects on them.

STEP 1 Cut enough snapped headers to be able to work uninterrupted on the panel.

STEP 2 Cut the mitred headers that will form the corners of the panel. These can be marked by first cutting a 65 mm square from one of the headers and placing on the floor in a builder's square.

STEP 3 Then, with a header placed alongside, a line can be scribed through the diagonals of the 65 mm cut and marked on the header. Repeat with the other header and then, marking back 5 mm from each line on either header, the two mitres can be cut. This will form an accurate 10 mm joint at 45° through both the headers.

STEP 4 Build in the cut bricks at either edge of the panel, making sure that the vertical bricks are laid to the correct gauge.

STEP 5 Mark the gauge for the headers along the bottom and then lay the first course to the line.

STEP 6 Mark the vertical gauge for the surrounding headers on either side of the panel.

PRACTICAL TIP

A piece of timber weighted across the top of the panel can be used to suspend vertical lines to ensure that the vertical headers are laid in the correct position. These can be built in stages or as work progresses.

STEP 7 Mark the correct gauge on the edge of the panel in order to lay the first bricks in the panel.

PRACTICAL TIP

Dry bond the middle brick in the basket weave and then the upright bricks can be laid in with a boat level to check for upright.

STEP 8 Repeat the process until the correct height of the basket weave bricks has been reached. Check continuously for alignment across the face.

PRACTICAL TIP

Whenever possible, stand back to look at your work as there may be an error that could be missed when working on your model in close proximity only.

STEP 9 Cut the corner mitres for the top two corners and correctly position.

STEP 10 Run in the top row of headers after checking for the correct gauge, which should correspond with the bottom course if the panel has been correctly built.

STEP 11 Finish with a half round joint, ensuring that the joint around the inner edge of the header surround is unbroken by any other joints.

PRACTICAL TASK

8. CONSTRUCT A HERRINGBONE PANEL

OBJECTIVE

To set out and construct a decorative herringbone panel.

The surround on this model is the same as for the basket weave panel. However, the herringbone panel itself is more complex and requires more care and attention when setting out. Nevertheless, if care is taken then the model should be fairly straightforward. As with the previous model, this panel is for purely decorative purposes and was often used in conjunction with other patterns such as basket weave on the same building. It would be fair to say that this was often a case of the building's owner as well as the bricklayer showing off, the former to display their wealth and the latter to show their level of skill in their trade.

TOOLS AND EQUIPMENT

Brick trowel

Lump hammer

Bolster chisel

Spirit level

Jointing iron

Builder's square

Scutch hammer

Steel tape measure

PPE

Ensure you select PPE appropriate to the job and site where you are working. Refer to the PPE section of Chapter 1.

Centre line

25 mm off centre line

Figure 7.60 Plan of herringbone panel

STEP 1 Cut enough snapped headers to be able to work uninterrupted on the panel.

Follow Steps 1–6 of the previous practical task.

STEP 2 Lay two bricks (Brick A and Brick B) at 90° on the floor and check with a builder's square.

STEP 3 Mark off the corner of Brick A to 215 mm along the length of Brick B. Now cut along the marks. Cut two more patterns like this.

PRACTICAL TIP

If the cut piece comes away from Brick A cleanly, it can be used in the middle of the pattern (Brick C). If not, another piece will have to be cut, making sure to allow for a 10 mm joint all around the cut.

STEP 4 Find the middle of the panel and set up cut piece from Brick C 25 mm from the left of centre. This will ensure that the herringbone pattern fits centrally.

STEP 5 Wall in the other two cut C pieces with their apexes 318 mm from the centre line of the panel and check for the correct level and angle.

STEP 6 Mark out the vertical (318 mm) and horizontal (52 mm) gauge.

STEP 7 Use vertical lines suspended from a piece of timber and horizontal lines to check for alignment in both directions.

PRACTICAL TIP

Make sure that all cut bricks are cut to maintain the correct 10 mm joints at the correct angles. Any irregularities will catch the eye on completion and detract from the panel's appearance.

STEP 8 Lay the bricks in for the herringbone pattern, continuously checking for alignment and 90°. The vertical lines should help prevent the pattern 'wandering'.

STEP 9 When the height of the infill has been achieved, mark the cut headers for the surround and lay to the line to complete the panel.

PRACTICAL TIP

A set square or piece of timber with a 90° angle can be used to check for correct positioning.

STEP 10 Finish with a half round joint, ensuring that the joint around the inner edge of the header surround is unbroken by any other joints.

9. CONSTRUCT DENTIL AND DOG TOOTHING OVER-SAILING COURSES

OBJECTIVE

To add decorative dentil and dog toothing features over arches.

These methods of laying bricks are primarily decorative features. Different combinations can be used and are often found around the top of a structure before the roofline commences. Nevertheless, care should be taken not to overdo over-sailing courses, otherwise they can look too clumsy and their decorative characteristics will be compromised. In the workshop, they are ideal as a feature over the arches as illustrated in Fig 7.61. Therefore, this practical exercise will assume that they are being built over some existing masonry.

TOOLS AND EQUIPMENT

Walling trowel	Jointing iron
Lump hammer	Scutch hammer
Bolster chisel	Steel tape measure
Spirit level	Block/line and pins

PPE

Ensure you select PPE appropriate to the job and site where you are working. Refer to the PPE section of Chapter 1.

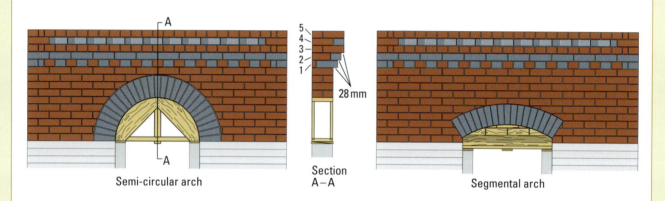

Semi-circular arch

Section A–A

28 mm

Segmental arch

Decorative features for both arches

Dentil work to course 1 of feature

Stretcher course to 2nd course of feature

Dog toothing to course 4 of feature

Header course to 3rd and top course of feature

Figure 7.61 Plan of dentil and dog toothing features to be added to arches

STEP 1 Position the first bricks of the over-sailing course at either end of the model, making sure that they are level from end to end as well as not tipping forward, which would give a drooping appearance to the overhanging bricks.

Figure 7.62 Measuring the over-sail course with a ruler

STEP 2 Add the closers, making sure that the cut face is not visible once the toothing has been formed.

PRACTICAL TIP

Place the bricks in by dry bonding to check that the pattern will work.

STEP 3 Run a line along the underside of the bricks to be placed and run the dentil course in using a tape or a cut piece of timber to maintain consistency when setting the recessed bricks.

STEP 4 Fill the spaces behind the recessed bricks with cuts.

Figure 7.63 Pointing the header course indents

STEP 5 Run the stretcher course in with the correct overhang and ensure that the back course and infill of cut closers is added so that the centre of gravity on the wall is maintained.

Figure 7.64 Running in the stretcher course (1)

Figure 7.65 Running in the stretcher course (2)

Figure 7.66 Running in the stretcher course (3)

STEP 6 Form the ends of the dog toothed course without an overhang and check for level before running in the back course.

STEP 7 Dry bond the dog toothed bricks to establish the pattern and cut the last brick to the right hand side. A line can then be attached to check for alignment and the bricks placed.

PRACTICAL TIP

The bricks should be at 45° to the face of the wall, meaning that the indents themselves will be at 90° to each other. Therefore a template can be cut from timber, so that the regularity of the dog toothings can be checked and maintained.

Figure 7.67 Checking the dog tooth angle

STEP 8 Point the toothings now as this will be difficult to complete once the next course is installed.

Figure 7.68 Pointing the toothings (1)

PRACTICAL TIP

Try to use just the right amount of mortar on the beds. You want to avoid too much squeezing out and dropping through to the dog toothing.

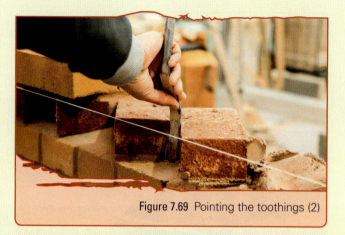

Figure 7.69 Pointing the toothings (2)

STEP 9 Build the header course corners without an over-sail and run in, then repeat for final course.

STEP 10 Fill the dog toothings with a fillet of mortar so that any water runs off the brickwork instead of settling on a ledge that would otherwise be present.

Figure 7.70 Making the fillet

STEP 11 Finish with a half round joint and brush to finish.

Figure 7.71 The finished model

TEST YOURSELF

1. 1Where would you find tumbling in?

   a. At the bottom of a buttress

   b. At the end of a wall, so that it can be extended

   c. In weather struck joints

   d. At the top of a buttress

2. If the radius on a ramp is large, what type of joints should be used?

   a. Flush joints

   b. 'V' joints

   c. Weather cut

   d. Curved

3. At what angle should the bed joints of buttresses be set to the slope?

   a. 30°

   b. 60°

   c. 90°

   d. 120°

4. Which of the following can impose tensile tension on walling?

   a. Wind on a long and exposed wall

   b. A wall that has earth piled up behind it

   c. A pillar that supports a heavy gate

   d. All of these

5. Which of the following types of reinforcement can be used for brick lintels, gate pillars and Quetta bonds?

   a. Mild steel rods

   b. Expanded metal

   c. Welded fabric

   d. Hoop iron

6. Herringbone bond features patterns of bricks that are laid at 90° to each other. How many degrees are they to the horizontal plane?

   a. 30

   b. 60

   c. 90

   d. 45

7. Instead of using single bricks a double herringbone bond uses which of the following?

   a. Double stretchers

   b. Bricks on end

   c. Diagonally cut bricks

   d. Horizontally cut bricks

8. When tumbling in courses are being constructed, what is the alternative name given to the template that can be cut from hardboard or plywood?

   a. Trammel

   b. Batten

   c. Gun

   d. Angle

9. One way of incorporating projecting bricks into a design is to cut the ends off so that they have a rough surface protruding. Which of the following is also **true**?

   a. The non-facing side of the brick also needs to be uneven

   b. The brick needs to be wedged in with additional reinforcement

   c. The edges of the brick need to be cut square

   d. The amount to which the bricks project is irrelevant

10. What type of course of bricks consists of a number of bricks being laid on end, side by side?

   a. Soldier

   b. Dentil

   c. Dog toothing

   d. Over-sailing

INDEX

ACKNOWLEDGEMENTS

The author and the publisher would also like to thank the following for permission to reproduce material:

Images and diagrams

Alamy: Adrian Sherratt: chapter 4 opener and 4.12, Building Image: 3.14, Colin Underhill: chapter 6 opener, Construction Photography: 4.9, EDIFICE: chapter 7 opener, Midland Aerial Pictures: 3.3, Peter Davey: chapter 1 opener, Photimageon: 7.10, ZUMA Press, Inc.: 2.1; **BSA:** 2.5; **2013 © Energy Saving Trust**: 3.16; **Fotolia:** 1.1, 1.2, 1.3, 1.5, 1.6, 1.7, 1.8, 1.14, 1.15, 1.16, 3.2, 3.8; **Helfen:** 2.4; **instant art:** table 1.15; **iStockphoto:** 1.11, 2.12, 2.13, 3.1, 3.4, 3.5, 3.10, 3.12, 3.13, 3.15, chapter 5 opener, 5.1; **Nelson Thornes:** 1.9, 1.10, 1.12, 1.13, 4.24, 4.25, 4.26, 4.27, 6.15, 6.16, 6.18, 6.19, 6.20, 6.21, 6.22, 6.23, 6.24, 6.40, 7.20, 7.21, 7.62, 7.63, 7.64, 7.65, 7.66, 7.67, 7.68, 7.69, 7.70, 7.71; **Peter Brett:** 2.3, 2.6, 2.7; **Science Photo Library:** Peter Gardiner: 1.4; **Shutterstock:** chapter 2 opener, chapter 3 opener, 3.6, 3.7, 3.9, 3.11.

Every effort has been made to trace the copyright holders but if any have been inadvertently overlooked the publisher will be pleased to make the necessary arrangements at the first opportunity.